Digital Infrastructures

An invisible network of digital technology systems underlies the highly visible networks of roads, rail, waterways, communication facilities, and power-lines. Increasingly, these systems are becoming the "infrastructure's infrastructure," providing crucial information and control systems to manage network demand, performance, reliability, and security. *Digital Infrastructures* presents an interdisciplinary analysis of the social, economic, organizational, and environmental forces shaping the information technologies (IT) that envelop these networks. Analyses of specific civil and environmental infrastructures are integrated with broader policy and management issues, including the recent concerns with using IT to manage these critical systems under crisis conditions.

The book covers three domains of digital infrastructures: foundations, cases, and crosscutting systems. First, core features are defined in terms of the technology, policy, and management dimensions associated with the application of IT to physical infrastructures. Second, indepth case analyses are presented for specific infrastructure sectors on the effect that information technologies are having on the functioning of civil and environmental infrastructures. Third, cross-cutting themes of economics, earth systems engineering, and international sustainability are applied to various infrastructures, with special attention to how these systems perspectives approach barriers to integrating information technology and infrastructure. A concluding chapter explores new directions and challenges being posed by issues such as security that are transforming the fundamental ways in which infrastructure is managed.

Digital Infrastructures is the first integrated treatment of how IT fundamentally affects the management of critical infrastructures. It provides the new infrastructure professional with state of the art concepts, methods and examples for use in creating public policies, strategic plans, and the design and implementation of new systems. It will be an essential book for upper level undergraduate and graduate courses in the management of critical infrastructure, environmental systems, and IT.

Rae Zimmerman is Professor of Planning and Public Administration and Director of the National Science Foundation-funded Institute for Civil Infrastructure Systems (ICIS) at New York University's Wagner Graduate School of Public Service, where she directs policy research in risk and security assessment, critical infrastructures, and environmental systems. **Thomas Horan** is Associate Professor and Director of the Claremont Information and Technology Institute at Claremont Graduate University's School of Information Science, where he leads national and regional studies on the use of information systems to guide public and private sector infrastructure development.

The Networked Cities Series

Series Editors:
Richard Hanley
New York City College of Technology, City University of New York, US
Steve Graham
GURU, Newcastle University, UK
Simon Marvin
SURF, Salford University, UK

From the earliest times, people settling in cities devised clever ways of moving things: the materials they needed to build shelters, the water and food they needed to survive, the tools they needed for their work, the armaments they needed for their protection – and ultimately, themselves. Twenty-first century urbanities are still moving things about, but now they employ networks to facilitate that movement – and the things they now move include electricity, capital, sounds, and images.

The Networked Cities Series has as its focus these local, global, physical, and virtual urban networks of movement. It is designed to offer scholars, practitioners, and decision makers studies on the ways cities, technologies, and multiple forms of urban movement intersect and create the contemporary urban environment.

Moving People, Goods and Information in the 21st Century
The Cutting-Edge Infrastructures of Networked Cities
Edited by Richard Hanley

Digital Infrastructures
Enabling Civil and Environmental Systems through Information Technology
Edited by Rae Zimmerman and Thomas Horan

Sustaining Urban Networks
The Social Diffusion of Large Technical Systems
Edited by Olivier Coutard, Richard Hanley and Rae Zimmerman

Digital Infrastructures

Enabling Civil and Environmental Systems through Information Technology

Edited by
Rae Zimmerman and Thomas Horan

Routledge
Taylor & Francis Group
LONDON AND NEW YORK

First published 2004
by Routledge
2 Park Square, Milton Park, Abingdon, Oxon OX14 4RN

Simultaneously published in the USA and Canada
by Routledge
270 Madison Ave, New York, NY 10016

Routledge is an imprint of the Taylor & Francis Group

Typeset in Times by Wearset Ltd, Boldon, Tyne and Wear
Printed and bound in Great Britain by TJ International, Padstow, Cornwall

British Library Cataloguing in Publication Data
A catalogue record for this book is available from the British Library

Library of Congress Cataloging in Publication Data
A catalog record for this book has been requested

ISBN 0-415-32460-2 (hb)
ISBN 0-415-32461-0 (pb)

To: Michael, Gabe, and Alexa
To: Rut, Ethan, and Kiley
For their patience and understanding

Contents

ACKNOWLEDGMENTS

The authors are indebted to the National Science Foundation, in particular to Dr Priscilla Nelson and Dr Joseph Bordogna for their vision and creativity in seeing the importance of this work. We also thank Dr Miriam Heller with whom we worked closely on its execution, and many of the other program directors at NSF who participated and lent the perspectives of their disciplines to this work. Grant support from the following NSF awards (Rae Zimmerman, Principal Investigator) provided invaluable resources for this work: The Institute for Civil Infrastructure Systems (ICIS), CMS 9728805; Bringing Information Technology to Infrastructure: A Workshop to Develop a Research Agenda, CMS-0091482; and Urban Research Services in a Time of Crisis: Lessons from September 11, CMS 0204660. Any opinions, findings, and conclusions or recommendations expressed in this book are those of the authors and do not necessarily reflect the views of the National Science Foundation.

We are especially indebted to Mara Cusker, a graduate research assistant and then assistant research scientist at ICIS. In addition, other graduate students and senior staff at ICIS, in particular, Nate Gilbertson and Stephen James are acknowledged for their very hard work in preparing the initial workshop report. Many thanks go to the numerous participants and presenters at the workshop for providing the spark and inspiration for this book. We particularly value the insights and assistance provided by Richard Hanley, the editor of this series, in reviewing the material. We are also grateful to the chapter contributors, who demonstrated a highly cooperative spirit; many attended group discussions on the book and all worked diligently to have their respective chapters fit within the overall goals of the volume. Finally, we want to thank Wendy Remington for her tireless assistance in compiling the manuscript.

Credits and Permissions

The authors and the publishers would like to thank the following individuals and institutions for giving permission to reproduce the material listed below. We have made every effort to acknowledge copyright holders, but if any errors have been made we will make the necessary corrections in subsequent printings.

Figure 1.3 Courtesy of the US Department of Transportation.
Figures 1.6, 1.7, 1.8 Courtesy of the National Science Foundation.
Figure 2.1 Courtesy of Ramesh L. Shrestha, University of Florida.
Figures 7.1, 7.2 Graphs were produced using the database provided courtesy of the North American Electric Reliability Council Disturbance Analysis Working Group.
Figure 7.3 Graph produced using data courtesy of The EPRI/DoD Complex Interactive Networks/Systems Initiative (CIN/SI).

Chapter 8
Portions of this chapter are reprinted with permission from Townsend, A. (2004) "Learning from September 11th: ICT infrastructure collapses in a 'global' cybercity," in Graham, S. (ed.) *The Cyber Cities Reader*, London: Routledge, 143–8.

Notes on Contributors

Braden Allenby is Professor at Arizona State University's Ira A. Fulton School of Engineering in the Department of Civil and Environmental Engineering. Prior to that he was Vice President, Environment, Health and Safety at AT&T, an adjunct professor at The University of Virginia's School of Engineering and at Princeton Theological Seminary, and the inaugural Batten Fellow at Darden Graduate School of Business at the University of Virginia. He is well known for his work in industrial ecology, and works with information systems and technology from an earth systems engineering and management perspective, studying the economic, environmental and social implications of technological systems, communications, infrastructure, and services. Dr Allenby has co-edited, authored and co-authored numerous textbooks in industrial ecology and systems engineering. He received his B.A. cum laude from Yale University, J.D. and Masters in Economics from the University of Virginia, and Masters and Ph.D. in Environmental Sciences from Rutgers.

Massoud Amin is Professor of Electrical and Computer Engineering, directs the Center for the Development of Technological Leadership, and holds the HW Sweatt Chair in Technological Leadership at the University of Minnesota. His research focuses on global transition dynamics for resilience and security of national critical infrastructures. He has over 20 years of security-related research and development experience. Before joining Minnesota in March 2003, he was with the Electric Power Research Institute (EPRI), where he coined the term "self-healing grid," led the development of over 24 technologies being transferred to industry, and after 9/11 directed all security-related research and development. He has led programs with military, governmental and private agencies, on national infrastructures, complex networks, adaptive intelligent controls, on-line decision making, system optimization and differential game theory. He is a member of numerous boards and committees, including the Board on Infrastructure and the Constructed Environment (BICE) (US National Academy of Engineering), and is a senior member of IEEE, AAAS, AIAA, and ASME. Dr Amin received his B.S. (cum laude) and M.S. degrees in electrical and computer engineering from the University of Massachusetts, Amherst and M.S. and D.Sc. degrees in systems science and mathematics from Washington University.

Timothy Beatley is Teresa Heinz Professor of Sustainable Communities in the Department of Urban and Environmental Planning, School of Architecture at the University of Virginia, where he has taught for the last eighteen years. His primary teaching and research interests are in environmental planning and policy, emphasizing coastal and natural hazards planning, environmental values and ethics, and biodiversity conservation. He has published extensively and his recent books are *Ethical Land Use* (Johns Hopkins University Press, 1994); *Habitat Conservation Planning: Endangered Species and Urban Growth* (University of Texas Press, 1994), *Natural Hazard Mitigation* (Island Press, 1999, with David Godschalk and others); and *An Introduction to Coastal Zone Management* (Island Press, 2002, Second Edition, with David Brower and Anna Schwab). In addition, he co-authored *The Ecology of Place* (Island Press, 1997, with Kristy Manning) on sustainability practices, *Green Urbanism: Learning from European Cities* (Island Press, 2000), and authored the forthcoming book *Native to Nowhere: Sustaining Home and Community in a Global Age*. He received his Bachelors in Urban Planning from the University of Virginia, MUP from the University of Oregon, M.A. Political Science and PhD in City and Regional Planning from the University of North Carolina at Chapel Hill.

Christopher Cluett is a Research Leader in the Transportation Division of Battelle in Seattle, Washington. He has 27 years of experience with Battelle in behavioral science research and project management, managing numerous Intelligent Transportation Systems (ITS) research and evaluation projects over the past decade. He recently led a research project for the Transportation Research Board (TRB) on the impact of information and communication technologies on state Departments of Transportation. Research areas include evaluation, institutional and societal analysis, public outreach and involvement, and focus group and survey implementation. He served from 1995 to 2000 as Chairman of the Societal Issues Task Force for ITS America and was a member of the Coordinating Council of that organization. He presents papers at national and international transportation meetings, and has published in a variety of professional journals. He is an active member of ITS America, the TRB, and the American Sociological Association. He is currently on the Board of ITS Washington, and is Co-Chair of the University of Washington's Department of Sociology Advisory Board. He received a B.A. from Williams College and an M.A. and Ph.D. in Sociology from the University of Washington.

Jesus Dumagan is Senior Economist with the Office of Policy Development (OPD), Economics and Statistics Administration, US Department of Commerce. His current research concentrates on the impacts of information technology on productivity and inflation, and appears in three US Department of Commerce reports. He was awarded (with collaborators) a Commerce Department Silver Medal for outstanding research on the transformative effects of information technology on the US economy. His current research focuses on decomposing aggregate productivity growth and overall inflation into component contributions (e.g. output and input categories). Publications appear in international academic journals (*Economics Letters*, *Economic Modeling*, *Energy Economics*, and *Resource and Energy Economics*) on superlative index formulas, computing cost-of-living indexes, modeling and estimation of consumer demand systems, measuring consumer welfare effects of environmental penalties and benefits from energy conservation, and analyzing trends in food-away-from-home consumption. He received his Ph.D. from the Department of Applied Economics and Management, Cornell University (Phi Kappa Phi), and was also a Fulbright-Hays Fellow at the University of Minnesota, where he earned an M.A. in Economics.

Jonathan Gifford is Associate Professor of Public Management and Policy in the School of Public Policy at George Mason University. Dr Gifford is a specialist on transportation policy. His expertise includes highway and urban transportation policy, aviation policy, and advanced technology in transportation systems. He recently completed *Flexible Urban Transportation* (Oxford: Pergamon, 2003), and is currently completing *The Interstate Highway System: A Policy Retrospective* (Cheltenham, UK: Edward Elgar, forthcoming). Dr Gifford directs and teaches in the Master's in Transportation Policy, Operations and Logistics program. He also teaches information technology, public policy analysis, and risk analysis in the Master of Public Administration program. Dr Gifford has worked in the US Office of Management and Budget, the Congressional Budget Office, and the Office of Technology Assessment. He is trained as a civil engineer and received his Master of Science and Ph.D. from the University of California, Berkeley, and his Bachelor of Science in Civil Engineering from Carnegie Mellon University.

David Henry is Senior Industry Analyst in the Office of Policy Development (OPD), Economics and Statistics Administration, US Department of Commerce, where he provides economic analyses for planned or existing policies on a wide variety of issues. He has co-authored the Department's *Digital Economy* reports over the past few years that address the economic conditions and importance of the IT sector of the economy. The reports estimate the share of IT industries in the economy, their contribution to US economic growth, effects on inflation, productivity, and the size and composition of the IT workforce. His work has extended to other issues including homeland security, crude oil price increases in response to supply disruptions, impacts of steel tariffs, and impacts of reducing emissions in connection with global climate change. Mr. Henry was the Chief Economist for the 1993 and 1995 Defense Base Realignment and Closure Commission. Prior to Federal service, Mr. Henry was the Senior Engineer for the National Association of Home Builder's Research Foundation, Inc. He received his B.S. in Forest

Science and Technology and M.S. in Civil Engineering from Pennsylvania State University.

Thomas Horan is Associate Professor, School of Information Science, Claremont Graduate University and Director of the Claremont Information and Technology Institute (CITI). Dr Horan's work investigates the management, social, and environmental implications of information technologies. In addition to his recent book, *Digital Places*, he has authored numerous technical articles and book chapters on digital technologies, some of which have appeared in *Communications of the ACM*, *Information Systems Frontiers*, *Journal of Urban Technology,* and *Handbook of Public Information Systems.* Dr Horan has led a variety of academic and applied research studies on the deployment of advanced technologies in local, regional, national and international settings. His research has been funded by the US Department of Transportation, the National Science Foundation, and the Fletcher Jones Foundation and has been used to devise private sector telecommunications systems, local e-government programs, and national technology policies. He is a member of the Academy of Management, the Association of Information Science, and the Association of Public Policy Analysis and Management. Dr Horan received his B.A. (Phi Beta Kappa) from the University of Vermont, his M.A. in Public Policy and Ph.D. in Public Affairs and Organizational Psychology from the Claremont Graduate University.

Mitchell Moss is Professor and Director of the Taub Urban Research Center at NYU's Wagner Graduate School of Public Service. Dr Moss teaches and conducts research on urban planning and development, and has written extensively on telecommunications and the future of cities. His research has been supported by the National Science Foundation, American Express Foundation, Andrew Mellon Foundation, and Charles H. Revson Foundation. Professor Moss is currently doing research on the way in which the September 11 attack on the World Trade Center has shaped planning and politics in the New York Metropolitan Region. His essays have appeared in the *New York Times*, *New York Post*, *New York Daily News*, and the *New York Observer.* Dr Moss received his B.A from Northwestern University in Political Science, his M.A. from the University of Washington in Political Science, and his Ph.D. in Urban Studies from the University of Southern California.

William O'Brien is an Assistant Professor in the Construction Engineering and Project Management program in the Department of Civil Engineering at the University of Texas at Austin. He focuses his professional efforts on improving collaboration and coordination among firms in design and construction. Dr O'Brien is an expert on construction supply chain management and electronic collaboration. He is especially interested in the use of the information technologies to support multi-firm coordination, and works with leading firms to implement web-tools to support practice. Previously, he was CSR/Rinker Assistant Professor in the M.E. Rinker, Sr. School of Building

Construction and Assistant Professor in the Department of Civil & Coastal Engineering at the University of Florida. Prior to returning to academia, Dr O'Brien led product development and planning efforts at Collaborative Structures, a Boston based Internet start-up focused on serving the construction industry. Dr O'Brien received a B.S. degree in Civil Engineering from Columbia University, an M.S. degree in Civil Engineering, an M.S. degree in Engineering-Economic Systems, and a Ph.D. from Stanford University.

Carlos Restrepo is a Ph.D. candidate at NYU's Wagner Graduate School of Public Service. His dissertation topic is the association between asthma and air pollution in New York City. He is a graduate research assistant at the Institute for Civil Infrastructure Systems (ICIS) and Project Manager for the ICIS component of the South Bronx Environmental Health and Policy Study, funded by the US Environmental Protection Agency. This is a joint project between ICIS, NYU's Nelson Institute of Environmental Medicine, and four South Bronx community groups that relates transportation and waste management activities in the South Bronx, New York, to environmental quality and public health issues. His publications and conference papers, include co-author of the NSF funded workshop report, "Bringing Information Technology to Infrastructure" and co-author of a chapter on the NSF funded workshop on the September 11 attacks in *Beyond September 11th* (U. of Colorado, Boulder, 2003). Prior to joining ICIS, he was a research assistant in an NSF-funded research project on performance measures and infrastructure. Before coming to NYU, Carlos worked for three years in El Salvador, where he is from, as a policy analyst for FUSADES, a non-profit organization. He received his B.S. in Engineering Physics from Lehigh University and his M.S. in International Development and Appropriate Technology from the University of Pennsylvania.

Lucio Soibelman is an Associate Professor at the Civil and Environmental Engineering Department at Carnegie Mellon University where he teaches courses relating to construction management, construction information systems, and Civil Engineering Artificial Intelligence. The general direction of Dr Soibelman's research has been the organization of diverse data into forms that are amenable to the application of Knowledge Discovery methods. His research in Knowledge Discovery concentrates on studying the increasing amount of available data, developing Data Mining and Knowledge Discovery in Databases methods, processes, and tools to generate novel knowledge from large construction databases, and organizing the large amount of available construction data with the development of improved information retrieval methods. Dr Soibelman received B.S. and M.S. degrees in Civil Engineering from Universidade Federal do Rio Grande do Sul (UFRGS), Brazil and his Ph.D. in Civil Engineering Systems from the Massachusetts Institute of Technology (MIT) and worked from 1998 to 2004 as an Assistant Professor in the Civil and Environmental Engineering Department of the University of Illinois at Urbana Champaign.

Roy Sparrow is Professor of Public and Nonprofit Management at the Robert F. Wagner Graduate School of Public Service, New York University. He previously served as Acting Dean, Associate Dean and head of the management program at the Wagner School. He specializes in public transportation policy and management with a particular interest in research on strategic change in organizations that design and deliver urban infrastructure services. Since 1998, he has served as Director of the Education Program of the Institute for Civil Infrastructure Systems (ICIS), a multi-university NSF-funded center for collaborative, interdisciplinary activities supporting infrastructure research, education, and outreach. He teaches graduate courses in Managing Public Service Organizations, Strategic Management, and Developing Management Skills. He received his B.A. from University of North Carolina, Chapel Hill and Ph.D. in Political Science from the University of California, Los Angeles.

Anthony Townsend is a Research Scientist and Lecturer of Public Administration at the Taub Urban Research Center at New York University. Dr Townsend has served as a technology consultant to major information technology companies including Nortel Networks, Quova, and Telegeography, Inc and has developed relationships with research groups at Ericsson, Intel Corporation, and AT&T. Prior to graduate school, Anthony worked as a technical support manager for several years at a regional Internet Service Provider in New Jersey and AT&T WorldNet. He is the author of over a dozen scholarly articles and book chapters on the impacts of new information and communications technologies on urban and regional development. His articles have also appeared widely in the trade press, including *Computerworld*, *On* (Corporate magazine of Ericsson), and *Wired News*. He received his Masters of Urban Planning from New York University and a Ph.D. from the Massachusetts Institute of Technology.

Rae Zimmerman is Professor of Planning and Public Administration at New York University's Robert F. Wagner Graduate School of Public Service. Since 1998, she has been Director of the Institute for Civil Infrastructure Systems (ICIS), a multi-university National Science Foundation (NSF)-funded center for collaborative, interdisciplinary activities on infrastructure research, education, and outreach. ICIS is a partner in the Homeland Security Center for Risk and Economic Modeling of Terrorism Events with the University of Southern California, and her work focuses on critical infrastructure. She authored *Governmental Management of Chemical Risk* (Lewis/CRC), co-produced *Beyond September 11th* (Boulder, CO, University of Colorado, 2003), and is a co-editor of *Sustaining Urban Networks* (Routledge, 2004). She has authored numerous papers on how interdependencies among transportation, water, and related environmental protection infrastructure impact communities served by this infrastructure. She has been an expert for government and industry on large infrastructure projects especially in the context of extreme events, and has directed over a dozen infrastructure and environmental research projects with funding primarily from NSF and the US Environmental Protection Agency (EPA). She is a Fellow and Former President of the Society for Risk Analysis, a Fellow of the American Association for the Advancement of Science, and has served on government and academic boards on the social and environmental dimensions of infrastructure. Prior to NYU, she was with the US EPA. Dr Zimmerman received her B.A. in Chemistry from University of California (Berkeley), Master of Urban Planning from the University of Pennsylvania, and Ph.D. in Urban Planning from Columbia University.

Part I

Overview

CHAPTER ONE

What are Digital Infrastructures?

Rae Zimmerman and Thomas Horan

Introduction

In a world that continues to increase in size and complexity, the dependence on information technologies (IT) that drive our life support systems is growing rapidly. Few other technologies have spread as rapidly. This book addresses the pervasive influence that IT has had on infrastructure, namely transportation, water supply and wastewater management, energy, and telecommunications, and its users. This is especially timely in light of the growing need for critical policy, management and technological choices about the reliability and security of IT and infrastructure systems, and in particular what was deemed critical infrastructure by the President's Commission on Critical Infrastructure Protection in 1997 (US Department of Commerce, Critical Infrastructure Assurance Office 1997) and again in 2003 by the White House (White House 2003).

As a point of departure, it is clear that digital technologies have enabled our core infrastructures to expand geographically and in terms of functional capability; for example, our transportation systems are now managed through regional information systems that can monitor conditions on different roadway systems and across transportation modes. Indeed, our cities and towns are now fundamentally dependent on technology to provide a range of monitoring, diagnostic, and control information that allows our society to function smoothly with little or no interruption and under a wide range of conditions. Users increasingly have enhanced options to interact with infrastructure, as a result of the expanded infrastructure capabilities IT can provide.

But, as Edward Tenner (1996) bluntly put it: "things bite back." For infrastructure managers, the implication is this: when IT is incorporated into the design of infrastructure systems for high performance, it can beget new problems with infrastructures and create unintended consequences.

The technology sword indeed has two edges: one edge facilitates unprecedented infrastructure performance and the other edge allows for dramatic infrastructure failure when environmental and/or technological conditions run afoul. This second edge of the sword was felt in 2003 when history repeated itself with a number of electric power blackouts around the world of magnitudes rarely seen before – in the United States, Canada, the United Kingdom, Italy, Sweden and Denmark. In a number of those cases, cyber failures were not or have not yet been identified as initiators. But the blackout of August 14, 2003, that affected about 50 million people across the northeastern and mid-western US and Canada making it the largest blackout, did have cyber origins in the form of software and computer failures according to the US-Canada Power System Outage Task Force (2004: 51–55).

The story does not stop there. A growing literature on the interdependencies among different kinds of infrastructure (see for example O'Rourke 1993; Rinaldi *et al.* 2001) underscores the fact that what starts out as an IT failure can quickly propagate throughout other systems. The causes can be accidents, natural hazards, or acts of terrorism, and IT failures range from very localized breaches of transmission lines, such as fiber optic cable, to more extensive main computer server outages. A common example of a localized breach is damage to fiber optic cable by construction crews, called "backhoe fading," which has been considered the largest source of public telephone network failures (Schneider 1999: 37). One such incident where an electrical power failure caused a breakdown of a phone system ultimately resulted in a shutdown of one of the three New York region airports in Newark, NJ (Hevesi 1991: A1). An example of a more extensive, and potentially very damaging IT failure is a breakdown of portions of the UUNet (Unix to Unix Network) backbone (UUNET Technologies, Inc., Fairfax, VA) that is central to the functioning of the Internet. One such failure occurred when nine out of 13 Domain Name System (DNS) root servers "responsible for helping to resolve domain names to their respective IPs" failed for an hour on October 22, 2002; another occurred in 1997 when seven of the servers were disabled (Internet Traffic Report 2002).[1] IT failures such as these can have a demonstrable impact on the real-time operation and long-term viability of civil and environmental infrastructures as they become increasingly dependent upon IT.

In addition to accidental disruptions, such as blackouts, acts of terrorism that target cyber infrastructure or indirectly damage it, have yielded even more pronounced impacts on infrastructure because of its interdependencies with IT. The September 11, 2001 attacks underscored the fact that while infrastructure may not be the direct target, indirectly the damages to IT facilities housed in destroyed buildings, and destruction or overloading of communication systems had widespread effects, for example, on vital communication capabilities to allow transportation and financial institutions to function (Zimmerman 2003) and on domain name servers as far away as South Africa (National Research Council 2002b).

Interdependencies are economic and social as well as physical. IT potentially primes the economic engine of infrastructure, which in turn affects the rest of the economy. The social implications of the growing dependency of society in general on IT, the "digital divide," have commonly been portrayed in terms of access to computers (US Department of Commerce 2000), and some have carried this further, pointing out that the cultural and social contexts of computer usage and educational support are equally important (Warschauer 2003). Consumers of infrastructure services are increasingly faced with IT-embedded service usage and payment systems where knowledge and proficiency with these systems is required to access infrastructure services.

Thus, the enabling of infrastructure by IT encompasses a very wide range of issues that need to be confronted to reinforce the positive features of IT and overcome the negative ones. It is for this reason that IT is identified as an important area for attention in terms of protecting and enhancing "critical infrastructures" in the US as well as internationally (Wenger *et al.* 2002).

Origins of this volume

This book had its origins in a workshop funded by the National Science Foundation (NSF) in June 2001, entitled, "Bringing Information Technology to Infrastructure: A Workshop to Develop a Research Agenda."[2] The purpose of the workshop was to identify research ideas for information technology and infrastructure around the themes of technology, management, and policy (Zimmerman *et al.* 2002). The workshop white papers and discussions revealed significant issues in each of these areas, as well as a host of crosscutting themes. In technology, new approaches to infrastructure system sensing, monitoring and control were outlined, including the use of new models to support the diagnosis of infrastructure systems for condition and vulnerability assessments. From the management discussions, it became clear that IT was shaping infrastructure organizations and organizational processes and had the potential to break down jurisdictional barriers among different kinds of infrastructure systems. In terms of policy, a host of federal, state and local policy directions were viewed as seriously affecting the nature, timing and use of IT in infrastructure systems. Some of the research needs identified at the workshop were issues regarding differential access by different populations and users of infrastructure to IT, security and privacy concerns, and the need for computer and analytical tools to support policy development.

While the workshop identified critical areas for attention, participants noted the paucity of integrated treatments of the IT and infrastructure intersection. This volume is aimed at providing such an integrated treatment by including both cross-infrastructure analyses of IT issues as well as indepth case analyses of IT uses in specific industries, recognizing that specific cases often have very widespread implications in spite of very specific, localized origins. Moreover, this volume strives to balance technical dimensions with social, organizational, and policy considerations. This integration can be seen in the conceptual framework that provides a foundation for the chapters.

A Conceptual Framework

This volume has a consistent underlying orientation regarding the role of digital technologies in civil and environmental infrastructures. The Management chapter (Chapter 3) explicitly addresses this orientation in the form of a "Conceptual Model for Managing Digital Infrastructures" (see Figure 3.1). While Chapter 3 describes the model in detail, it is worth introducing its components here as it represents a useful schema for considering the broader domain of digital infrastructures. The key consideration is that digital infrastructure systems are viewed as having several interweaving socio-technical dimensions. The first such dimension is the physical infrastructure itself (e.g. water, transport, electricity, telecommunications). Upon this platform rests the IT dimension that contains electronic means by which the infrastructure can be sensed, monitored, analyzed, and controlled. Surrounding the IT and physical layers are various users and institutional and policy actors. Users interact with the infrastructure, based on their demands for services and information they may need about the infrastructure services (e.g. price, quality, availability). In addition to end-users are a variety of public and private institutions that plan, manage, regulate, operate, and otherwise service the infrastructure and its digital features. Finally, these institutions are enmeshed in social, economic, policy and physical contexts that set various boundary conditions on the infrastructure, its users, and managing organizations. In many ways, this volume is a journey into how these various components interact with each other across infrastructures and under varying conditions of deployment and performance.

Information Technology and Infrastructure Overview

Definitions

This volume uses the term information technologies (IT) broadly to encompass information, communications and computing functions. Another term, information and communication technology (ICT) is also commonly used to portray information technology and the communication component, and the sectors that comprise these activities as defined by the OECD in 1998 and revisited in 2002.[3] The chapters in this book generally use IT as the term of choice though it can be considered to encompass ICT. Certain communications functions are often collectively referred to as telecommunications.[4] Computing infrastructure has been subdivided further into the following components: "the Internet, embedded/real-time computing (e.g. avionics systems for aircraft control, Supervisory Control and Data Acquisition (SCADA) systems controlling electrical energy distribution), and dedicated computing devices (e.g. desktop computers)" (NRC 2002b: 135). Finally, digital technologies is another term used to characterize portions of the information technology sector. Chapter 2 addresses the full range of sensing, monitoring, control and communications functionalities that many of these systems encompass. Chapter 9 gives the economic sectors in the US economy that typically comprise information technologies.

The Expanse of Infrastructure and its Growth Trends
Both traditional infrastructure and IT have grown dramatically over the past century and are expected to continue to be robust in the twenty-first century. They often share the same customers or markets as well as the underlying machinery providing the services and the sites where these services are provided.

Major infrastructure systems continue to link cities and regions throughout the world with massive distribution systems that now require IT for access to information and services from remote locations. In the US for example there are about four million miles of federal and state roadways, about 100,000 miles of rail lines for the largest of the systems (US Department of Commerce, Critical Infrastructure Assurance Office 1997: A-11), about one million miles of major water lines, hundreds of thousands of miles of oil and gas lines (NRC 2002a) and in excess of 200,000 miles of electric power transmission lines that provide 230,000 volts or more (US-Canada Power System Outage Task Force 2003: 3). These are a testament to the ability of infrastructure to transgress large spaces, yet can IT keep up?

Trends in infrastructure usage, many of which are described in other chapters, underscore its dramatic growth over the past few decades, and the task at hand for IT. Changes that occurred within the US exemplify the steady if not explosive use of infrastructure:

- By 2001, total energy consumption was three times what it was in 1949 (Figure 1.1); electricity use grew by a factor of thirteen over that same period (US Department of Energy, Energy Information Administration (EIA) 2001) as shown in Figure 1.2. When growth rates for population and the gross domestic product (GDP) are taken into account, the trend is less dramatic but still sizable. Using US Department of Commerce figures, between 1949 and 2001, population grew by 91 percent (from 149 million to 285 million) (US Department of Commerce, Bureau of the Census 2003: Table B-34, 317) and real GDP by 600 percent, which translates into a growth of 57 percent in per capita energy consumption and a decline from 20.6 thousand BTUs per dollar of real GDP to 10.6 thousand by 2002 (US Department of Commerce undated web site: Table 1.1.6).

- Most measures of automobile travel show a steady rise over the twentieth century. Growth in vehicle miles of travel, for example, approximates a linear trend over the decade of the 1990s (see Figure 1.3 and Chapter 6).

- Public transit usage as shown in Figure 1.4 increased by 8.2 percent from 1998 to 2002 (based on data from the American Public Transportation Association (APTA) 2002). Chapter 6 in this volume points out that the highest level of public transit travel occurred in 1946. Ridership declined through the 1970s, then increased, and stabilized in the early part of the twenty-first century.

- Water usage between 1950 and 1995 rose by 123 percent (see Figure 1.5), and population rose during that period by 72.6 percent (US Bureau of the Census 2000). The details of this trend are provided in Chapter 5.

One important factor in being able to keep up with this demand is the service and design life of facilities. Though estimates are highly variable depending on facility usage, materials, and condition standards, design lifetimes for

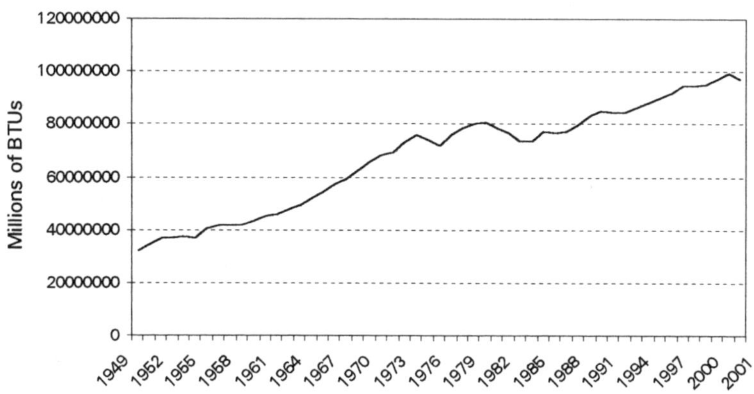

1.1 *Total Energy Consumption, United States, 1949–2001.*
Source: Graphed from data provided in Energy Information Administration (EIA) US Department of Energy (2001) *Annual Energy Review 2001*, Energy Perspectives: Trends and Milestones 1949–2001, Table 1.3, Online. Available at: http://www.eia.doe.gov/emeu/aer/txt/ptb0103.html.

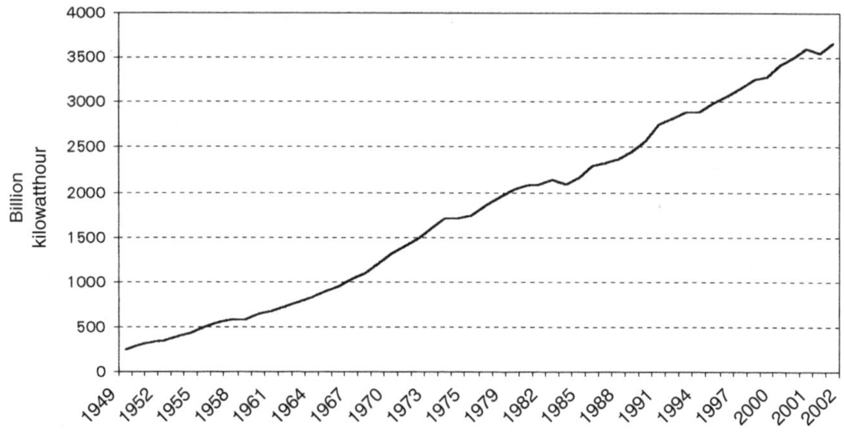

1.2 *Electricity Use, United States, 1949–2002.*
Source: Graphed from data provided in Energy Information Administration (EIA) US Department of Energy (2001) *Annual Energy Review 2001*, Energy Perspectives: Trends and Milestones 1949–2001, Table 8.1, Online. Available at: http://www.eia.doe.gov/emeu/aer/txt/ptb0801.html.

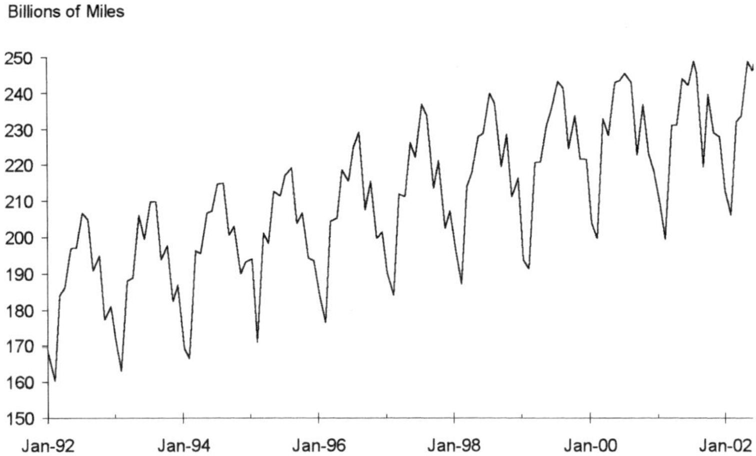

1.3 Vehicle Miles of Travel, United States, 1992–2002.
Source: US Department of Transportation, Federal Highway Administration, Office of Highway Policy Information, Online. Available at: http://www.fhwa.dot. gov/ohim/tvtw/tvtpage.htm. Courtesy of the US Department of Transportation.

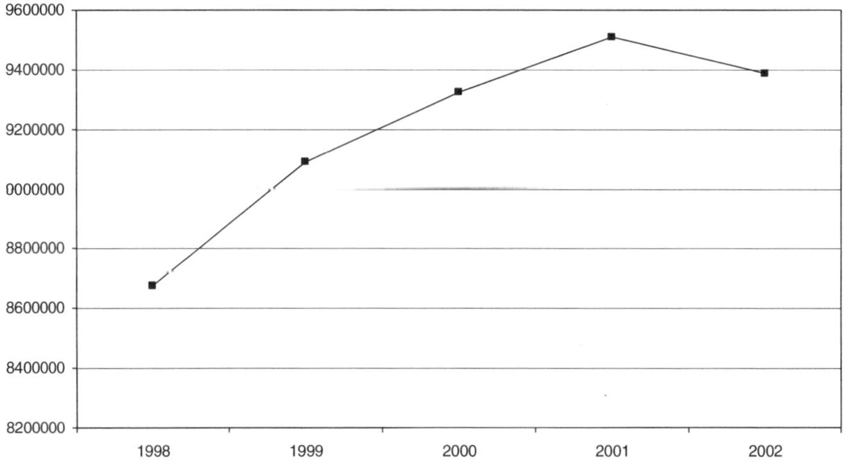

1.4 Transit Trips, United States, 1998–2002.
Source: Graphed using annual totals from monthly data from the American Public Transportation Association, "APTA Ridership Reports Statistics–United States Mode Totals; APTA Quarterly Transit Ridership Reports", Online. Available at: http://www.apta.com/research/stats/ridershp/#A1, "Transit Agency Data," "APTA Ridership Reports," "Mode Totals."

Notes

a Transit passenger trips include heavy rail, light rail, commuter rail, trolleybuses, buses, aerial tramways, automated guideways, cable cars, ferryboats, inclined planes, monorails and vanpools.

b APTA defines unlinked passenger trips as: "the number of passengers who board public transportation vehicles. Passengers are counted each time they board vehicles no matter how many vehicles they use to travel from their origin to their destination." See: http://www.apta.com/research/stats/ridershp/definitions.cfm.

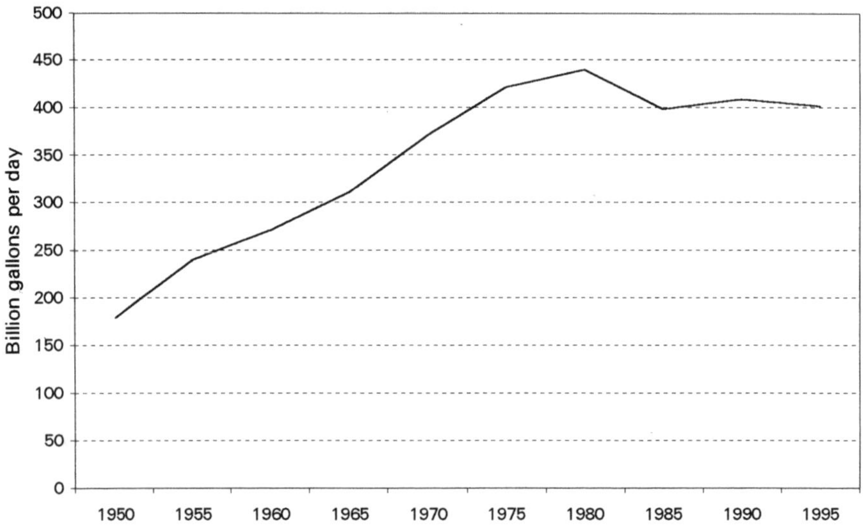

1.5 *Trends in Water Use, United States, 1950–1995 (latest dates available).*
Source: Graphed from figures provided in US Geological Survey (USGS). Online. Available at: http://ga.water.usgs.gov/edu/tables/totrendbar.html (accessed December 5, 2003) and Solley, W.B., Pierce, R.R. and Perlman, H.A. (1998) *Estimated Use of Water in the United States in 1995*, Denver, CO: US Geological Survey, Table 1, p. 7.

infrastructure are at least a decade and often a century or more. Some examples of these design lifetimes are given below:

- *Roads:* 20 years, but depends on condition standard, type of traffic, and roadway materials (Marland and Weinberg 1988: 328)
- *Bridges:* variable, hundreds of years (Marland and Weinberg 1988: 328)
- *Buses:* 8–15 years (Armstrong-Wright 1986: 32)
- *Light rail:* 25 years (Armstrong-Wright 1986: 32)
- *Rapid rail:* 30 years (Armstrong-Wright 1986: 32)
- *Power plants:* historically, 25–30 years; a 70-year design standard is used in Russia (Marland and Weinberg 1988: 328)
- *Dams:* variable, hundreds of years (Marland and Weinberg 1988: 325)
- *Water and wastewater treatment and distribution systems:* 15–100 years, depending on the component (US EPA 2002: 11, Table 2-1, see compilation in Chapter 5).

These trends and conditions highlight the significant changes associated with the upgrading (including technological upgrading) of our infrastructures in the developed regions of the world. As illustrated in Chapter 10, the situation in the developing world is equally if not more daunting. Throughout Asia, Africa, and Latin America, local governments are struggling to provide basic infrastructure for economic and community growth. The challenge in this case is to conceive of a means for the innovative (e.g. low-cost) use of technology as part of the general investment strategy for improving infrastructures in these regions.

IT Growth and Technological Changes

A key factor affecting the integration of IT into infrastructure (whether in developing or developed regions) is a set of trends that make IT more affordable, reliable, and powerful than could have been imagined a few short decades ago. These conditions include the rapid growth in product use and technological changes in terms of the exponential growth in processing power and IT capacity.

Product Use and Byproducts

IT products and services are numerous and highly diverse. The use of some of the more common ones show dramatically high growth rates.

- Mobile cell phone use continues to rise. In 1990, the OECD reported a total of 10,537,234 mobile cellular subscribers among Organisation for Economic Co-operation and Development (OECD) countries with about half (5,283,055) in the US By mid-1999 this number had grown to 292,780,578 (OECD 2000: 73). This amounts to a penetration rate on average for all OECD countries from one per 100 inhabitants in 1990 to 26.8 by June 1999 (National Science Board 2002: Appendix Table 8-3), as shown in Figure 1.6.

- Internet use continues to rise. Over a four-year period between 1997 and 2001 the number of Internet hosts that house, for example web sites, quadrupled in the US (see Figure 1.7). Access to the Internet continues to grow and by the end of 2002, 60 percent of Americans were reported to have Internet access (Horrigan and Rainie 2002).

- High-speed broadband use continues to rise. By 2003, 16 percent of Americans had access to high-speed broadband service, a dramatic rise in those with access since 2000. In other countries, like South Korea and Canada, the rise in access is reported to be even higher with over half of the households with high speed broadband connections (Horrigan 2003: 1–2).

Not only has growth in products been widespread, but the growth in waste accumulated has also occurred, reflecting product expansion. According to the National Safety Council (Itasca, IL), "500 million US computers are estimated to become obsolete by 2007" (Irrinki 2000: 33). Assuming 200 million phones in

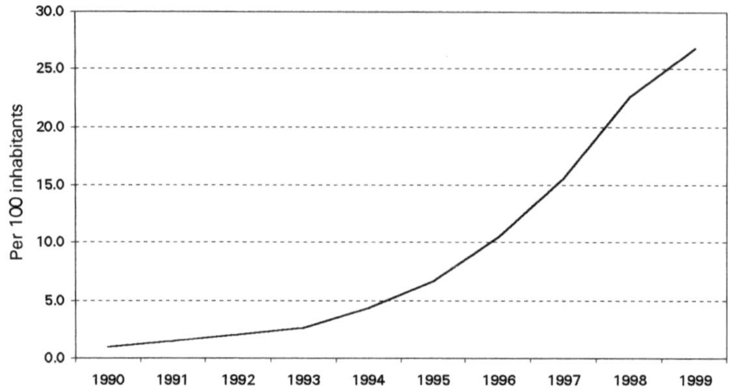

1.6 Average Mobile Phone Use Penetration, OECD Countries, 1990–1999.
Source: Graphed from National Science Board (NSB) (2002) *Science & Engineering Indicators – 2002*, Volume 2: Appendix Tables, Arlington, VA: National Science Foundation (NSB–02–01). Appendix Table 8.3, Page A8–3, Online. Available at: http://www.nsf.gov/sbe/srs/seind02/append/c8/at08-03.pdf. Courtesy of the National Science Foundation. A similar graph appears in Volume 1, Figure 8–5, p. 8–8 as growth in mobile phone subscribers. Data source in NSB tables is: Organisation for Economic Co-operation and Development (OECD) (2000) Cellular Mobile Pricing Structures and Trends, Paris.

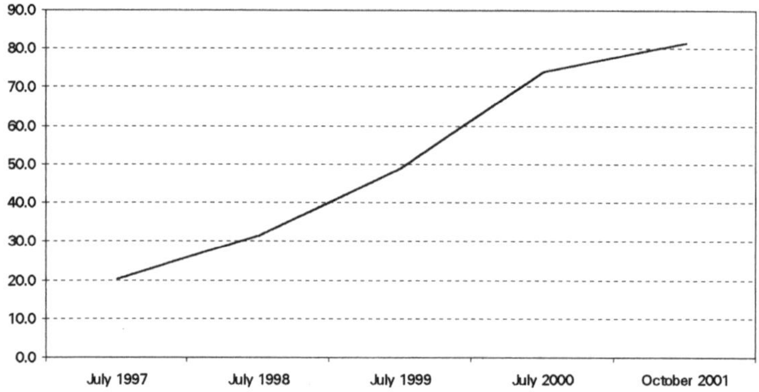

1.7 Average Internet Hosts per 1,000 Inhabitants, OECD, 1997–2001.
Source: Graphed from National Science Board (NSB) (2002) *Science & Engineering Indicators – 2002*, Volume 2: Appendix Tables, Arlington, VA: National Science Foundation (NSB–02–01). Appendix Table 8.4, Page A8–4, Online. Available at: http://www.nsf.gov/sbe/srs/seind02/append/c8/at08-04.pdf. Courtesy of the National Science Foundation. Data source in NSB tables is: Organisation for Economic Co-operation and Development (OECD) (2000) Cellular Mobile Pricing Structures and Trends, Paris. Data from Netsizer http://www.netsizer.com.

circulation for 1.5 years each, Fishbein has estimated that about 130 million phones would be retired per year (Fishbein 2002: 22). Beyond concerns over waste accumulation are the broader social conditions associated with environmental influences of products and materials across the entire sequence of IT activities (Richards *et al.* 2001). Concerns about the toxic materials in IT products and equipment have identified nanotechnologies as a source of environmental pollution, which has initiated major governmental research on the topic (Masciangioli and Zhang 2003: 106A–7A).

Technological Changes

Technological changes in the equipment used for IT have enabled the speeds for data processing to increase, and hence the capabilities for infrastructure. These changes have depended upon breakthroughs in the composition of various processing components expressed primarily in terms of chip capacity. Chip capacity, in terms of the density of transistors on a chip, continues to grow. "The number of transistors on a chip has doubled approximately every 12 to 18 months for the past 30 years – a trend known as Moore's Law ... named for Gordon Moore of Intel, who first observed it" (National Science Board 2002: A8-1). This rate of growth is shown in Figure 1.8. "The number of transistors per square inch on integrated circuits had doubled every year since the integrated circuit was invented" (Webopedia 2001). The National Research Council (1998: 9) notes further that, "Combined with numerous other technological advances, this capability has led to a doubling of microprocessor power every 18 months." Because of the limits imposed by silicon-based film, it was thought that chip density was approaching a limit, however, recent advances in nanotechnology may enable Moore's Law to continue to operate (Keren *et al.* 2003).

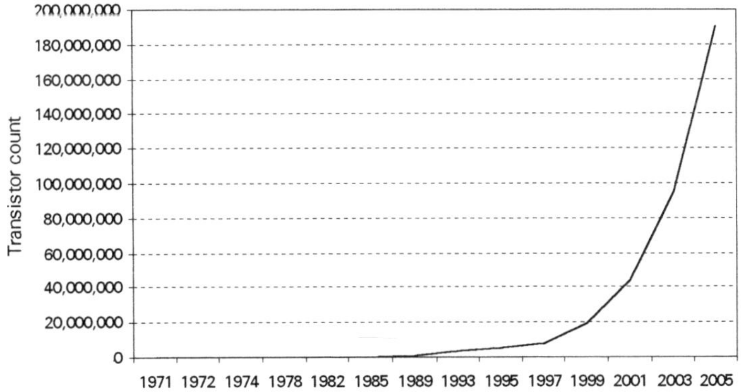

1.8 *Moore's Law, 1971–2005.*
Source: Graphed from National Science Board (NSB) (2002) *Science & Engineering Indicators – 2002*, Volume 2; Appendix Tables, Arlington, VA: National Science Foundation (NSB–02–1), Appendix Table 8.1, Page A8–1, Online. Available at: http://www.nsf.gov/sbe/srs/seind02/append/c8/at08–01.pdf. Courtesy of the National Science Foundation. A similar graph appears in Volume 1, Figure 8–1, p. 8–5 (log scale). Data sources in NSB tables is: 1971–01 – Available at http://www.intel.com/pressroom/kits/quickrefyr.htm; 2003–05 – International Technology Roadmap for Semiconductors. 2000 International Technology Roadmap for Semiconductors, available at http://public.itrs.net/Files/2000UpdateFinal/ORTC2000final.pdf.

Data storage mediums are also changing rapidly. Over a period of two decades or so, storage has changed from magnetic tape, to diskettes, to CDs, to DVDs, with associated hardware changes. Programs and associated software have changed rapidly as well.

All of these changes have created a situation where infrastructure systems have had to constantly adapt and retrofit IT to their operations at a rate that is typically faster than the usual time period for rehabilitating infrastructure. These differentials raise questions about the potential for increasing very formidable cascading failures due to increased infrastructure interdependencies discussed earlier in this chapter.

Integration
As alluded to in our conceptual model for managing digital infrastructure, these rapid technology developments and the potential for increasing cascading effects need to be considered within the systems domain of infrastructure and its users. While the technological advancements are impressive, the most pressing challenge is the need to harness IT in a manner that produces social, economic and community gain from infrastructure systems. As a society we have come to value the day-to-day service levels of a modern infrastructure. The roads we travel upon, the water we drink, the phone calls we make, the Internet we surf, the lights we turn on – all of these functions have become increasingly reliant on reliable IT systems. To this end, this book is about harnessing the power of technology to deliver these day-to-day services in a socially responsive, economically efficient, environmentally responsible and, yes, secure manner. While aware that things do indeed "bite back," the premise of this book is that technology can be harnessed to benefit our quality of life.

Organization of this Volume

Part I of the book gives three key dimensions of the IT and infrastructure issue, namely, technology, management, and policy. As the workshop noted above featured, it is in these contexts that IT and infrastructure co-exist and are constrained or enabled. Chapter 2 opens with technology, since in order to make informed decisions about new infrastructure systems, professionals and managers need to understand the fundamentals of the technology. Chapter 3 follows with the management of the technology, since the reconciliation of infrastructure complications, technological developments and policy choices and tradeoffs arrives at the doorstep of infrastructure managers. This chapter both provides a conceptual model for managing digital infrastructures and grounds this model in a series of case analyses from different infrastructures in diverse parts of the world. The last chapter in this section, Chapter 4, deals with policy. The infrastructure manager operates in a policy environment, which can condition, impinge or facilitate the decision making process. This chapter analyzes the policy environment, noting the state of policy with respect to integrating IT and infrastructure.

Part II provides an in-depth examination of the experiences within specific infrastructure sectors in incorporating IT – water, transportation, electric power, and telecommunications – to learn how those infrastructure systems are adapting to IT, and whether the different sectors share common experiences.

Part III enlarges the perspective on the IT/Infrastructure issue with four cross cutting analyses. The first analysis considers the economic implications of the convergence of IT and infrastructure, providing an empirical foundation for the investment context surrounding digital infrastructures. The second analysis considers the global context for digital infrastructures, with a focus on the emerging conditions in developing regions of the world. The third provides extensive examples and insights in the use of IT to enable new green technologies for infrastructure services within buildings. The fourth analysis places digital infrastructures in a broad environmental and philosophical perspective, suggesting how technology can play an integral role in devising environmentally responsive programs and policies. The concluding chapter takes up additional cross-cutting issues identified in the earlier chapters and related literature, and provides new directions. These themes and directions include the importance of understanding digital infrastructure interdependencies, the need to learn from extreme events involving digital infrastructures, the challenge of designing digital infrastructures that can be understood and used by diverse end-users and organizations, educational needs to support all of this, and, finally, the need to consider digital infrastructures as part of a global concern for development and sustainability.

Throughout these sections, the chapter authors strive to balance the general principles of IT and infrastructure with practical experiences and innovations. While the general tone does lean toward the optimistic, each chapter confronts the complexities of the subject matter including the many crises and breakdowns that can confront civil and environmental infrastructures. Society has made great strides in its use of IT, and this volume focuses on extending that achievement to the area of civil and environmental services.

Notes

1 The Domain Name System (DNS) is "a distributed database that keeps the name-to-address mappings for the Internet" (National Research Council 2002b: 4-1).

2 The workshop, was held in June 2001 in Arlington, VA by the Institute for Civil Infrastructure Systems (ICIS) (www.nyu.edu/icis) located at New York University's Wagner Graduate School of Public Service, which is a National Science Foundation (NSF) funded consortium of four universities: NYU, Cornell University, Polytechnic University of NY, and the University of Southern California. The workshop brought together about fifty participants from government, industry, and academia who are specialists in IT, infrastructure or both. Short papers prepared by the participants and keynote addresses by Professor William Mitchell of MIT, Dr Joseph Bordogna, Deputy Director of the National Science Foundation, and Dr Braden Allenby' Vice President of Environment, Health and Safety of AT&T. The full report of the workshop was prepared by Zimmerman, Gilbertson and Restrepo (2002) and is located at: www.icisnyu.org/admin/files/FinalITreport3.pdf.

3 Organisation for Economic Co-operation and Development (OECD) (2002) "Measuring the Information Economy 2002," pp. 81–3, Online. Available at: www.oecd.org/dataoecd/34/37/2771153.pdf (accessed January 16, 2004).

4 Telecommunications is defined in the American Heritage Dictionary (2000) as "the science and technology of communication at a distance by electronic transmission of impulses, as by telegraph, cable, telephone, radio, or television" and as "the electronic systems used in transmitting messages, as by telegraph, cable, telephone, radio, or television," and to connote "a message so transmitted." The American Heritage Dictionary of the English Language, Fourth Edition (2000) Boston, MA: Houghton Mifflin Company. dictionary.com, Online. Available at: dictionary.reference.com/search?q=telecommunications (accessed December 10, 2003).

References

American Heritage Dictionary of the English Language, Fourth Edition (2000) Boston, MA: Houghton Mifflin Company, dictionary.com, Online. Available at: http://dictionary.reference.com/search?q=telecommunications (accessed December 10, 2003).

American Public Transportation Association (2002) *APTA Quarterly Transit Ridership Report*, Online. Available at: http://www.apta.com/research/stats/ridershp/ A1 (accessed January 16, 2004).

Armstrong-Wright, A. (1986) *Urban transit systems: Guidelines for examining options*, World Bank Technical Paper Number 52: Urban transit series, Washington, D.C.: World Bank.

Fishbein, B. (2002) *Waste in the Wireless World: The Challenge of Cell Phones*, New York, NY: Inform.

Hevesi, D. (1991) "Phone Breakdown in New York City Snarls Air Traffic: Long Distance Affected," *New York Times*, September 18, p. A1.

Horrigan, J.B. (2003) "Adoption of Broadband to the Home," *Pew Internet & American Life Project Memo*, Washington, D.C.: Pew Internet & American Life Project, May, Online. Available at: www.pewinternet.org/ (accessed January 16, 2004).

Horrigan, J.B. and Rainie, L. (2002) "Counting on the Internet," Washington, D.C.: Pew Internet & American Life Project, December 29, Online. Available at: www.pewinternet.org/ (accessed January 16, 2004).

Internet Traffic Report (2002) "Backbone DDoS," Online. Available at: www.internettrafficreport.com/event/2.htm (accessed January 15, 2004).

Irrinki, S. (2000) "The Digital Utility. Information technology enhances performance," *Water Environment & Technology*, December, pp. 29–33.

Keren, K., Berman, R.S., Buchstab, E., Sivan, U. and Braun, E. (2003) "DNA-Templated Carbon Nanotube Field-Effect Transistor," *Science*, Vol. 302: 1310, November 21.

Marland, G. and Weinberg, A.M. (1988) "Longevity of Infrastructure," in Ausubel, J.H. and Herman, R. (eds) *Cities and Their Vital Systems*, Washington, D.C.: National Academy Press.

Masciangioli, T. and Zhang, W-X. (2003) "Environmental Technologies at the Nanoscale," *Environmental Science & Technology*, March, 1 pp. 103A–8A.

National Research Council (NRC) (1998) *Fostering Research on the Economic and Social Impacts of Information Technology*, Washington, D.C.: National Academy Press.

National Research Council (NRC) (2002a) *Making the Nation Safer: The Role of Science and Technology in Countering Terrorism*, Washington, D.C.: National Academy Press.

National Research Council (NRC) (2002b) *The Internet Under Crisis Conditions: Learning from September 11*, Washington, D.C.: National Academy Press.

National Science Board (2002) *Science & Engineering Indicators – 2002*. Vol. 2: Appendix Tables, Arlington, VA: National Science Foundation (NSB-02-1), Online. Available at: www.nsf.gov/sbe/srs/seind02 (accessed January 16, 2004).

Organisation for Economic Co-operation and Development (OECD) (2000) *Cellular Mobile Pricing Structures and Trends*, Paris, Online. Available at: www.oecd.org/dataoecd/54/42/2538118.pdf (accessed January 16, 2004).

Organisation for Economic Co-operation and Development (OECD) (2002) "Measuring the Information Economy 2002," Online. Available at: www.oecd.org/dataoecd/34/37/2771153.pdf. (accessed January 16, 2004).

O'Rourke, T.D. (1993) "Prospectus for Lifelines and Infrastructure Research," in Stenquist, B. (ed.) *The Art and Science of Structural Engineering; Proceedings of the Symposium Honoring William J. Hal*, Upper Saddle River, NJ: Prentice-Hall, Inc., pp. 37–58.

Richards, D.J., Allenby, B.R. and Compton, W.D. (ed.) (2001) *Information Systems and the Environment*, Washington, D.C.: National Academy Press.

Rinaldi, S.M., Peerenboom, J.P. and Kelly, T.K. (2001) "Identifying, Understanding and Analyzing Critical Infrastructure Interdependencies," in *IEEE Control Systems Magazine*, December, pp. 11–25.

Schneider, F.B. (ed.) (1999) *Trust in Cyberspace*, Washington, D.C.: National Academy Press.

Solley, W.B., Pierce, R.R. and Perlman, H.A. (1998) *Estimated Use of Water in the United States in 1995*, Denver, CO: US Geological Survey.

Tenner, E. (1996) *Why things bite back: Technology and the revenge of unintended consequences*, New York: Alfred A. Knopf.

US Bureau of the Census, Population Estimates Program, Population Division (2000) "Historical National Population Estimates: July 1, 1900 to July 1, 1999." Online. Available at: eire.census.gov/popest/archives/pre1980/popclockest.txt. Internet Release Date: April 11, 2000; Revised date: June 28, 2000 (accessed January 16, 2004).

US-Canada Power System Outage Task Force (2004) *Final Report on the August 14th 2003 Blackout in the United States and Canada: Causes and Recommendations*, The Task Force, April.

US Department of Commerce (2000) *Falling Through the Net: Toward Digital Inclusion. A Report on Americans' Access to Technology Tools*, Washington, D.C.

US Department of Commerce, Bureau of the Census (2003) *Economic Report of the President*, February, Table B-34, Population by Age Group, 1929–2002, p. 317.

US Department of Commerce, Bureau of Economic Analysis, *Table 1.1.6, Real Gross Domestic Product, chained dollars, interactive data for National Income and Product Accounts*, Online. Available at: www.bea.gov (accessed January 16, 2004).

US Department of Commerce, Critical Infrastructure Assurance Office (CIAO) (1997) "Critical Foundations: Protecting America's Infrastructures, The Report of the President's Commission on Critical Infrastructure Protection," Washington, D.C.: CIAO, Online. Available at: www.ciao.gov/resource/pccip/PCCIP_Report.pdf (accessed January 16, 2004).

US Department of Energy, Energy Information Administration (EIA), US Department of Energy (2001) *Annual Energy Review 2001*, Energy Perspectives: Trends and Milestones 1949–2001.

US Department of Energy, Energy Information Administration (2002) *Monthly Energy Review*, October, Online. Available at: www.eia.doe.gov/mer (accessed January 16, 2004).

US Department of Transportation, Federal Highway Administration, Office of Highway Policy Information, Online. Available at: www.fhwa.dot.gov/ohim/tvtw/tvtpage.htm (accessed January 16, 2004).

US Environmental Protection Agency, Office of Water (2002) *The Clean Water and Drinking Water Infrastructure Gap Analysis*. EPA-816-R-02-020. September.

US Geological Survey (USGS) Online. Available at: ga.water.usgs.gov/edu/tables/totrendbar.html (accessed December 5, 2003).

UUNET Technologies, Inc., Fairfax, VA, Online. Available at: www.uunet.net (accessed January 16, 2004).

Warschauer, M. (2003) "Demystifying the Digital Divide," *Scientific American*, Vol. 289, No. 2, 42–9, August.

Webopedia (2001) "Moore's Law." Online. Available at: webopedia.internet.com/TERM/M/Moores_Law.html (accessed January 16, 2004).

Wenger, A., Metzger, J. and Dunn, M. (2002) *International Critical Information Infrastructure Handbook*, Zurich, Switzerland: Swiss Federal Institute of Technology, Center for Security Studies and Conflict Research.

White House (2003) "2003 Homeland Security Presidential Directive/Hspd–7," Washington, D.C. The White House. Available online at: http://www.whitehouse.gov/news/releases/2003. Accessed July 2, 2004. December 17.

Zimmerman, R. (2003) "Public Infrastructure Service Flexibility for Response and Recovery in the September 11th, 2001 Attacks at the World Trade Center," in Natural Hazards Research & Applications Information Center, Public Entity Risk Institute, and Institute for Civil Infrastructure Systems, *Beyond September 11th: An Account of Post-Disaster Research*. Special Publication #39. Boulder, CO: University of Colorado, pp. 241–68.

Zimmerman, R., Gilbertson, N. and Restrepo, C. (2002) Bringing Information Technology to Infrastructure: A Workshop to Develop A Research Agenda. Final Report. New York, NY: NYU-Wagner/Institute for Civil Infrastructure. National Science Foundation grant, CMS-0091482, Online. Available at: www.icisnyu.org/admin/files/FinalITreport3.pdf (accessed January 16, 2004).

CHAPTER TWO

Technology and Engineering Dimensions: Collecting and Interpreting New Information for Civil and Environmental Infrastructure Management

William O'Brien and Lucio Soibelman

Introduction

The digital infrastructure – the interconnected web of sensors, actuators, wired and wireless communication networks, and computer systems that support civil and environmental systems – is growing to be as complex and multi-faceted as the physical infrastructure it is meant to serve. Ensuring the quality and reliability of the digital infrastructure is increasingly a mission critical task for agencies such as utilities and departments of transportation. Operating agencies cannot ignore or postpone indefinitely deployment of the digital infrastructure. There are immediate imperatives to monitor and detect terrorist actions such as chemical and bioweapon attacks. Similarly, population growth drives increased demand on the physical infrastructure for example, in the form of increased capacity utilization. Changing expectations and quality of life also place demand on infrastructure. Coupled with a decrease in investment, there is a need for new methods to make do with current facilities. The digital infrastructure promises to monitor the condition of our physical infrastructure, providing the information to make more effective use of our current resources while meeting user needs. Power generation and distribution may be better balanced across the power grid via real-time load monitoring, avoiding potential failures and price spikes from peak loads (Rabaey *et al.* 2001). Intelligent transportation systems promise to help reroute traffic to smooth and speed flows on congested city streets (McQueen and McQueen 1999).

Imperatives for security and the ability to improve operating efficiency, and meet increasing demand, are not the sole drivers for the digital infrastructure.

The availability of economical, non-destructive sensors combined with low-cost communication networks such as the Internet and wireless devices are driving new visions from smart structures to smart cities. Smart structures embed sensors in building materials to monitor performance and direct corrective action, whether to identify necessary maintenance or to react to severe structural conditions such as hurricanes and earthquakes (Srinivasan and McFarland 2000). Smart buildings network a variety of sensors within a building, enabling improvements in indoor air quality while saving energy and allowing local configuration of environmental conditions (Loftness *et al.* 2001). Smart cities refers to deploying sensors across the built infrastructure and connecting them via the Internet to provide real-time data about building conditions, location and status of operations and emergency personnel, traffic conditions, etc. This comprehensive data set is then used by computer models to determine the best response to emergencies, provide directions for traffic control, and generally improve the level of intelligence about a given city's infrastructure (Roush 2001).

Visions of the scale of smart cities demonstrate the promise of the digital infrastructure. Unfortunately, many visions face significant technical challenges. It is the purpose of this chapter to provide a conceptual framework for understanding these challenges and associated opportunities. As the number of technologies and applications for the digital infrastructure is enormous, the authors do not attempt to provide a comprehensive survey. Rather, the authors suggest a number of challenges: (1) development and deployment of sensors and related data gathering and control devices; (2) modeling and interpretation of streams of data from multiple, distributed sensors; (3) integration of data from sensors with other information pertinent to management of the physical infrastructure; and (4) development of appropriate methods to present users with the myriad of information about the digital infrastructure. Each challenge is briefly detailed in the next two sections. The first section surveys existing and emerging sensor technologies for monitoring the physical infrastructure and discusses the modeling challenges presented by distributed sensors. The second section describes the broader information environment of the digital infrastructure, including integration and presentation of information for the many users of information. Some brief closing comments are made concerning implementation of the digital infrastructure in the context of the broader information environment and technical challenges.

Technologies for Monitoring the Physical Infrastructure

Several sensor technologies exist or are in development that promise to radically improve the speed, precision, and range of data available about the physical infrastructure. The number of technologies and potential applications is too numerous to enumerate. Suffice it to say that almost any data currently being collected or that could be posited for collection has a corresponding sensor technology. Hence, rather than attempt an exhaustive review of technologies, the authors have found that a useful categorization is to divide the sensor technologies into those that are permanently tied to a specific infrastructure and those that are mobile. Some specific examples of permanent and mobile sensors are detailed below to provide an overview of the challenges and opportunities offered by new sensor technologies. Future trends are also briefly detailed. A salient feature of almost all the new sensor technologies is their capabilities for non-destructive evaluation, particularly with regard to monitoring constructed facilities. Large-scale, real-time instrumentation is now feasible, providing both researchers and operations personnel enormous amounts of data. In turn, this presents significant modeling issues to combine data from multiple sensors into an overall performance assessment. This section concludes with a discussion of sensor data modeling.

Types of Sensor Deployment

Permanently Deployed Sensors

Fiber optic sensor systems are an example of emerging, permanently deployed sensors (CII 2003). Fiber optic sensors are the hair-thin cables used in the telecommunications industry except they are wrapped around or embedded in concrete structures. Data from such systems can be downloaded on-site or transmitted to a remote location. Fiber-optic sensors can monitor shrinkage, creep, and other processes that cause weakening in high-performance concrete structures. Fiber optic sensing is a natural outgrowth of aerospace research, because of its use in monitoring aeronautical and space structures composed of advanced materials. Sensors enable precise strength measurement as the concrete cures, as pieces are trucked to the site and placed, and during the first year of use. Fiber-optic sensors further provide information on the effects of stress long before signs of fatigue begin to show visibly, allowing engineers to address potential problems before they become serious.

Fiber optic sensor systems offer many advantages over traditional methods that rely largely on visual inspections. Benefits include:

- less time and expense than visual inspections
- improved accuracy due to detection of internal stresses
- real-time, remote monitoring of structural response
- availability of performance data throughout the facility lifecycle, improving construction methods and identifying maintenance issues
- increased systematization to enable comparison of data sets from different sensors.

The relatively low cost and continuous assessment capabilities provided by fiber optics provide other benefits to the research community. Long-term performance of advanced composite materials can be compared to conventional materials if both are used in a structure (e.g. a bridge uses a few high-performance concrete girders along with traditional concrete). Comparative evaluation is also possible among novel materials designed to withstand heavier loads and last longer than ordinary concrete.

Fiber optic systems are an example of sensors that must be put in place during construction of the facility. However, there is a need for sensors that can be deployed on the existing infrastructure, whether buildings, electrical systems, or water supplies. Several such technologies are in development, ranging from chemical detectors wired to a central monitoring station to more complex assessment technologies. An example of an assessment technology is Continuous Acoustic Monitoring for Structures (Elliott 1999). Acoustic energy (sound waves) is released when, for example, a wire breaks. Sensors mounted on the structure record this energy and transmit data to an on-site data acquisition and signal-processing unit. Recorded data is downloaded periodically to a processing center, where proprietary software is used to analyze and classify events. Reports are issued tabulating the time and location of probable wire breaks. This information helps structural engineers and their clients to ensure long-term structural integrity through periodic replacement of broken tendons. Localized areas of high failure rates may indicate structural inadequacy, requiring further investigation and possible repair. Further, the ability to calculate the frequency of wire breaks allows statistical prediction of future failure rates as well as assessment of the extent of past failures. Statistical prediction also allows accurate budgeting for future repairs.

Mobile Sensors

Sensor technologies such as fiber optics and acoustic monitoring have the advantage of providing detailed, continuous assessment of the performance of the physical infrastructure. Further, as costs for sensors decrease and as new facilities are increasingly deploying some types of sensor systems (e.g. environmental controls for heating, ventilation, and air conditions (HVAC) systems are increasingly centralized and electronically monitored), the data available to monitor the physical infrastructure will increase in quality and quantity. However, by their nature, permanently deployed sensors are limited in the type of data they can collect as well as the physical area surveyed. Hence, there is a corresponding need for mobile sensors. Mobile sensors include those that can be moved to a location for scanning from a fixed point and those that include automotive capabilities.

Examples of automotive mobile sensors are Smart Sewer Assessment Systems. Smart Sewer Assessment Systems include robots equipped with a variety of sensors (optical, ultrasonic, microwave, and magnetic, as well as a global position system) and typically a monitoring station. Examples include the KARO (Kuntze *et al.* 1995), Smart Pigs (Willke 1998), and PIRAT (Campbell *et al.* 1995) systems. The Smart Pigs system, for example, can accurately detect, size, and locate corrosion or any other anomalies in pipelines. Once the problem is detected the information can be used to develop a pipeline de-rating schedule, implement a repair or replacement program, determine if re-inspection is necessary, and evaluate effectiveness of a corrosion inhibitor program. The need for such capabilities is growing as existing sewer systems age. Examples of problems are dramatic collapses of pipelines from groundwater infiltration, and exfiltration of sewage into the groundwater and the surrounding soil causing dramatic environmental pollution. Germany, for example, pays an estimated 51 billion Euros in annual restoration of failed sewer lines (Kuntze *et al.* 1995). In the US, the 1996 Clean Water Needs Survey estimates that the cost of sewer infiltration/inflow correction and replacement and rehabilitation for 2001 would have been US$11.6 billion (1996 dollars adjusted to 2001 dollars) (US EPA 2002: 20). Hence, accurate preventive assessment of sewer conditions can provide significant economic benefits.

In contrast to automotive sensors, several technologies exist that can be transported and scan a fixed area. One such technology is Acoustic Wave Sensors to test indoor air quality. Modern office buildings are generally considered safe and healthful working environments. However, energy conservation measures have minimized the infiltration of outside air and contributed to the buildup of indoor air contaminants. Investigations of indoor air quality often fail to identify any harmful levels of specific toxic substances (CII 2003). The efforts to pursue stronger and more efficient systems for sensing the environmental contamination have been highly focused since September 11, 2001.

A variety of portable environmental monitoring systems have been designed, built, and field-tested using Surface Acoustic Wave (SAW) (AWST 2003) devices to provide rapid, reversible, sensitive, and quantitative detection of individual volatile organic compounds (VOCs), like airborne chemical agents that may be emitted by utilities. SAW devices are coated with polymers that absorb VOCs. Two independent responses of the SAW sensor (wave velocity and attenuation) are measured to provide information about the chemical species absorbed by the coating. The changes in the wave velocity and attenuation occur because the film coating the sensor softens and becomes heavier when it absorbs the contaminant. Sensing is rapid and reversible, and coatings can be optimized for particular chemical selectivity and sensitivity. SAW technology allows the manufacturing of a smaller, lighter, and most importantly, more chemical specific instruments. Available SAW equipment provides instruments and sensors for the detection of nerve agents, blister agents and several classes of toxic industrial chemicals that may be used in chemical warfare, terrorist incidents or hazardous material spills. Customers include federal agencies (special operations, protective services, explosive ordnance disposal) and municipal authorities (fire and police departments including bomb squads and emergency medical technicians).

3D Laser Scanning is an example of a technology that is mobile but can be deployed to scan from either a fixed point or attached to a moving object such as an airplane (Carter and Shrestha 1998; Hwang *et al.* 2003). 3D laser scanning has improved dramatically in recent years, allowing accurate measurements to a few millimeters (depending on the technologies employed). Several scanning technologies exist, but all work in the same manner. The laser scans a broad area, and

2.1 ALSM (Airborne Laser Swath Mapping) Scan for Damage Assessment of the World Trade Center Site Post 9/11. Courtesy of Ramesh L. Srestha, University of Florida.

each reflection is a dot used to build a point cloud representation of surface geometry of the subject. Four pieces of information are captured for each individual dot: the x, y, z coordinates and a return intensity, which they can use to map a color or grayscale over the cloud. The image that appears in the computer screen is a 3D cloud, which, depending on the scan density and technologies used, can capture minute details. These details allow assessment of tolerances during construction and can identify cracks and structural creep on existing facilities when compared to a reference model. An example of laser technology in practice is shown in Figure 2.1, a scan for damage assessment of the World Trade Center site, including damage to the telecommunications and transportation infrastructure. The scan allowed emergency personnel and structural and municipal engineers to rapidly assess damage and direct responses even though the site was covered with smoke.

Future Trends – From Health Monitoring to Health Saving

The sensors reviewed above can be classified as health monitoring technologies. Reports are made, for example, about the condition of existing facilities or presence of toxic chemicals. However, new technologies are in development that go beyond health monitoring to health saving. The health saving paradigm allows structures to autonomously take corrective action to structural distress. For example, self-healing concrete contains hollow adhesive-filled brittle fibers. Adhesives are released by failure of fibers when and where cracking occurs, and this results in a self-repair mechanism for the concrete (Dry 1994). Broader visions for smart structures promote the use of real-time monitoring of structural loads and stresses to allow the building to take corrective action (Srinivasan and McFarland 2000). This may include active reaction to dynamic loads from hurricanes or earthquakes or more mundane localized forces and failures. Actuators (devices for mechanical action in response to a signal) embedded in the structure will adjust local stresses. Piezoelectric and electrostrictive materials that respond mechanically to electrical fields are presently the most popular candidates for adaptive actuator systems because of their high dielectric constants, elastic modulus and high-frequency possibilities. Shape memory alloys (SMA) that regain their shape by heat after deformation also have many applications in composite structures due to their ability to induce large deformations and to be drawn into wires, which may be aligned with fibers in the host composite (Furuya 1995).

Nanotechnologies also promise significant contributions to both health monitoring and health saving approaches to maintain the physical infrastructure. Nanotechnology concerns products and devices on the minuscule scale of 1.0 to 0.01 of one nanometer (one-billionth of a meter). Presently, nanotechnology products are used in vehicle applications as impact detectors activating airbags and as nanocatalysts controlling air emissions. However, a recent workshop anticipates several applications to infrastructure (Livingston 2003): Nanosensors may be embedded in road pavement to monitor the processes that contribute to deterioration and cracking. The data would be accumulated in a database for researchers to use for extending the service lives of pavements. Similar sensors on bridges might monitor vibrations and loads, enabling researchers to assess structural weaknesses and conditions and fix them long before they are even apparent to human inspectors. With regard to health saving approaches, nanotechnologies could be used to develop self-healing materials composed of molecules that are able to rejoin themselves after being cut. One possible application would be roadside structures, such as guardrails, which would be capable of recovering rapidly if struck by a vehicle. Other applications for metal structures are nanolayer coatings that are self-cleaning and water-repellent to reduce corrosion and prolong the metal's life.

Integrated Modeling of Constructed Facilities

An important question related to sensors permanently tied to infrastructure objects is how to model the streaming data. Aktan *et al.* (2002) describe the design and implementation process for health monitoring on a major bridge. The data channels – sensor outputs – numbered 485 with capabilities to monitor traffic, weather, and structural response. Of these 485 data channels, 474 focused on monitoring the structural response of the bridge. As each channel provides real-time sensor data, it is clear that no human could effectively monitor the data to make intelligent decisions about structural performance and associated activities such as identifying maintenance needs. There is a need for computer tools to model the streaming data to provide useful assessments of performance. For their demonstration project, Aktan *et al.* (2002) describe a process of creating and calibrating a finite element analysis model (e.g. a structural analysis model composed of many small elements) with streams of sensor data to provide an integrated assessment model. Implementation challenges include accurate characterization of the physical structure, identification of phenomena affecting performance, location and calibration of sensors, data acquisition control and synchronization, data quality assurance, real-time information management, and data archiving. Most of these challenges are thorny but tractable. However, ensuring the correct characterization of the simulated model performance (finite element or other) in conjunction with sensor data remains a fundamental research challenge. There is little empirical history to build from, and models confined to simulation experiments may easily miss important assumptions and behaviors.

Hence, the availability of sensors to monitor a given component of the physical infrastructure is not a panacea. Comprehensive engineering approaches are needed to understand the interplay between sensor placement and data modeling to provide an integrated assessment. These approaches need to be grounded in research that examines the history of structural performance and sensor data. Historical trends and patterns may allow better interpretation of real-time data. Similarly, predictive models for a given structure may draw data from multiple sources to create an enlarged comparative basis for predictions. Such models will need to fuse data from several sources, each with likely different formats. As discussed in the next section, such integration is a non-trivial task. It will take time, experience, and research to make deployment of distributed sensors useful for large-scale modeling.

New Means of Communication

Building from the example above of placement and monitoring of sensors on a bridge, it is useful to comment on the physical means of communication between distributed sensors and data repositories. Information obtained through sensing technologies has traditionally been sent via wired communication, although there has been a trend from copper to fiber-optic cables. However, wired systems are being augmented and gradually replaced by wireless devices. Such devices include digital cellular systems and wireless local area networks for communications, and satellite systems for monitoring and positioning (e.g. Global Positioning Systems). Many sensors broadcast information over a wireless network, making them easier and less costly to deploy. At the same time, lack of a wire generally increases reliability as there are less physical systems to maintain. Similarly, devices such as radio-frequency identification (RFID) tags (Jaselskis 2003) provide a kind of wireless bar coding system. RFID tags are increasingly being used to track materials in logistics management applications, but have broad potential to provide information for infrastructure and emergency management (e.g. to broadcast information about a facility's physical composition to emergency responders). More broadly, the increasing capacity of wireless communication networks makes it easier to communicate rich data (e.g. convergence of voice and video with simpler text and numeric data). Wireless technologies promise to facilitate increasingly rich information sets in the digital infrastructure.

However, it is important to note that emerging communication technologies face several issues like the lack of reliability of the communication nodes. Further, communication standards are evolving for many systems, and several companies sell proprietary data collection systems for collecting and relaying data that is otherwise based on open standards. With the rapid evolution of technologies, there is high likelihood that the digital infrastructure will contain a panoply of wired and wireless communication systems. As discussed in the next section, such a variety of systems may hinder integration efforts, making use of the data challenging.

Making Sense of New Information: Users, Challenges, and Approaches

The previous section discussed current and emerging technologies for the digital infrastructure. Sensor and networking technologies provide information in real-time that heretofore has been practically unobtainable. However, the discussion above does not fully reflect the challenges and issues facing deployment of digital infrastructures. Consideration must be made for the wide range of potential users who may benefit from the information provided by new technologies. Integrating that information across types of users and presenting it in useful ways poses considerable technical challenges. This section discusses these challenges, first discussing the range of users and needs (the information environment) and then providing an overview of the technical approaches to support user needs. While discussion of human needs of course must also consider a variety of managerial, policy, and legal issues, the reader is directed to other chapters in this volume for coverage of these issues.

The Information Environment: Users and Challenges

The information environment for deployment of technologies to support the digital infrastructure is depicted in Figure 2.2. At the bottom of Figure 2.2, there are the various sensors deployed on the physical infrastructure (e.g. bridges, water supplies, etc.). Also at the bottom of the figure are external data sources that may contain information relevant to management of the physical infrastructure, including weather reports, building codes, news bulletins, hazardous materials information, etc. Immediate consumers of sensor information and information from external data sources are the various agencies responsible for the infrastructure such as departments of transportation and utilities, as well as owner/operators of major facilities (e.g. universities as large land owners have facilities that may contain hazardous materials, and often maintain water and

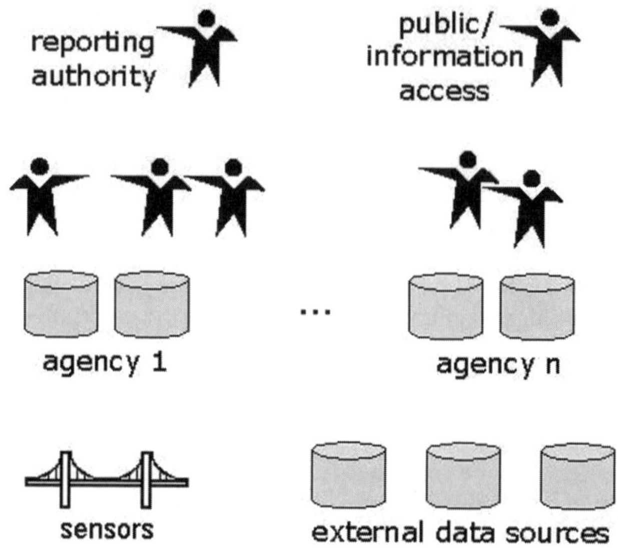

2.2 *The Information Environment for Infrastructure Management.*

electric utilities). Agencies such as fire and emergency services are also consumers of information about the state of the physical infrastructure and should be considered part of the information environment. All these agencies maintain internal databases and applications in addition to sensor information. For example, fire and emergency services agencies record data about hydrant location and property descriptions for their service region. Above the operating agencies are the reporting authorities that may coordinate or supervise local agencies (e.g. local, state, and national governments as well as agencies such as the Federal Emergency Management Agency (FEMA) that collect data from regional operating agencies and/or intervene in emergency situations). Rounding out the information environment is the public who require information about the physical infrastructure.

There are a large number of actors who comprise the information environment for the digital infrastructure. Consider as an example Gainesville, Florida, hometown of the University of Florida. Gainesville is a city of about 150,000 residents. Gainesville is the population center of Alachua County, Florida (although the county has several smaller, rapidly growing towns that assert a politically separate identity from Gainesville). Some of the most densely populated areas of Alachua are unincorporated portions adjoining Gainesville. Both the city and county have separate government structures, and each maintain their own fire and police departments in addition to administrative offices. There are some overlapping responsibilities; for example, Gainesville Regional Utilities (water and electric) serves the greater Gainesville region, including unincorporated regions of Alachua. However, many agency functions are duplicated between the city and county. Each agency has its own set of computer systems. In general these systems are not interoperable, although basic coordination of emergency services stems from a centralized Computer Aided Dispatching system. (As an example of current interoperability difficulties, information about residents is stored in the property appraisal and utility customer records databases. These are not searchable by fire rescue personnel from the field, and hence the rescue personnel have only a limited idea of the number of persons to search for in an emergency.) Coordination among utilities and emergency services is further complicated by the University of Florida, which maintains its own police force and has facilities for cogeneration of power and water reclamation. The University also has special infrastructure and facility needs, including labs with hazardous materials and an operating experimental nuclear reactor. Between the city, county and university, there are sixteen[1] operating agencies directly involved in maintenance of the physical and digital infrastructure or involved in response to a crisis in the physical infrastructure. These sixteen agencies are supplemented by dozens of other agencies and departments that use or supply information pertinent to maintenance of the physical infrastructure or emergency response. And these agencies also are subject to reporting authorities at the state and national level. The public is also very diverse, consisting of long-term residents, students, major professional offices in engineering and administrative services, four hospitals, light manufacturing facilities, two local television stations and several newspapers and radio stations. Each constituency requires different types of information and has different channels for receiving it. A relatively small city such as Gainesville has a rich and multifaceted information environment. Larger cities and regions are likely to have even more complex information environments that the digital infrastructure must serve.

The range of actors in the information environment – and associated range of computer systems and agency specific languages – poses considerable challenges for the realization of the digital infrastructure. Beyond the research and deployment issues for modeling the information provided by sensors (described above), there are challenges integrating distributed information and presenting information in a coherent and useful manner to individual users. The challenges of integration and presentation range from simple to complex. A simple application is the deployment of sensors to detect contaminants in water supplies. Sensor data must be integrated with existing water quality data sets to record changes and provide a history for comparative analysis. Beyond the effort of deploying the sensors and network they report on, integrating the information with a utility's existing data set/data systems is relatively straightforward and can likely be performed with in-house personnel.

However, fully realizing the power of the digital infrastructure requires more complex integration of data than simple deployment of sensors. Any time information is put to multiple uses or crosses organizational boundaries it is likely to increase complexity significantly. For example, Rabaey *et al.* (2001) envision the use of widely dispersed sensors and actuators across the power network (including sensors in buildings at point of electrical consumption) to radically improve the efficiency of power generation and distribution. Of course, such a vision requires integration of thousands of information systems across numerous agencies and facility owners. Fortunately, new sensors and information systems that are likely to conform to a pre-established data standard makes such a large-scale integration effort tractable.

Some of the most challenging integration efforts are likely to be at the local level across operating agencies in towns such as Gainesville. As noted above, a useful integration effort is to provide information about residents stored in utility databases to field rescue personnel. The smart city vision to network information systems together is broader, seeking to provide teams of responders real-time visual and textual information from sensors and databases. For example, when responding to a fire and potential chemical leak at the University of Florida, the rescue team could locate smoke and heat sources, track field personnel, call up building plans, access data about likely building occupants and hazardous materials, and communicate with utility personnel about water and electrical shutoffs. Other than sensors that must be deployed, most of this information is currently stored electronically. However, almost none of it is readily available to field rescue personnel with current systems. The basic networking infrastructure is in place as fire trucks and police cruisers are outfitted with laptop computers and networked to the dispatching system. What are missing are connections to the numerous computer systems that hold the data relevant to field responders. Consider that each of the sixteen core agencies in Gainesville has numerous computer systems (the fire department has at least four primary systems), some of them bought commercially and some of them home grown. Each of these computer systems is customized to the standard operating procedures of its primary agency. Loosely, each system will speak a different language or jargon. Putting in place an integrated query interface across even a subset of the relevant data sources is a formidable task.

Beyond the challenges of integrating information, there are issues in presenting it to users in an effective manner. Challenges vary intensely by type of application. For example, sensors on a bridge may sound an alarm of impending structural failure. Presentation of the warning information to a Department of Transportation (DOT) operations center is fairly straightforward. However, presentation of information that coordinates a response to that alarm is considerably more complex, relying on multiple data sources and involving multiple agencies. The combination of information from sensors and legacy (existing) data sources can overload users, and there is an associated need for methods to help humans parse the data. Several authors identify potential pitfalls. Norman (1998) notes that in the presence of too much information, humans can make poor choices because they identify the wrong information appropriate to the decision. Norman traces the evolution of cockpits in aircraft from a few controls to hundreds, noting that the trend in modern design is to reduce complexity. Norman's concerns are most pressing in situations when there is a need for quick response and complex data from multiple sources. Schneiderman (2002) more broadly calls for computing tools that promote universal usability, including interface designs that allow users to perform analysis and data manipulation necessary to explore alternatives and solve problems. Brown and Duguid (2000) reinforce Norman's and Schneiderman's critiques, further noting that a reliance on digital sources for problem solving can abstract away the social context the information was generated in. This is particularly problematic for coordination among teams composed of disparate individuals.

Returning to the example of emergency response, consider an integrated information environment where field personnel have access to building plans, hazardous materials data, location of field personnel, etc. How should this information be organized and presented to help, as opposed to hinder, operations? Referring to Brown and Duguid's social context, should the rescue personnel trust floor plans (that may be out of date) or notes from a fire rescue pre-plan (that may have occurred a year ago)? Beyond interpretation of the information in the digital infrastructure, responders must not become reliant on it as a sole source of information or judgment. Certainly there is the chance that the network will be unavailable, but a greater problem is missing or inaccurate information derived from the digital infrastructure. A fire-fighter shouldn't run a hose around a corner when there is a hydrant across the street (that wasn't marked in a database and hence not visible in the digital infrastructure). Presentation of information to aid human judgment is a significant problem that must be addressed to realize robust use of the digital infrastructure.

Technical Approaches to Information Management for the Digital Infrastructure

The information management challenges detailed above are not unique to the digital infrastructure, and a variety of approaches to address these challenges exist in the computing literature. Some of the main approaches are detailed below. These include information integration, including knowledge extraction and mediation, personalization and user modeling, and information security. It is important to note, however, that the problems posed by the information environment for digital infrastructures are complex, requiring robust solutions. While promising approaches exist, they generally represent partial solutions. The demands of the information environment will likely prove a fertile ground for research and development in academia and industry. At the same time, the approaches detailed below do hold promise for rapid application to deployment of the digital infrastructure. The authors especially note technical approaches likely to support implementation.

Information Integration

Information integration among the various computer systems that make up the information environment is an essential component of a fully realized digital infrastructure. As noted above, a fundamental problem is that the various agencies and actors in the information environment are likely to have computer systems with different means of representing information. The panoply of legacy systems poses a particularly pernicious problem as every implementation must be customized. Hammer and O'Brien (2004) and O'Brien *et al.* (2002) note that to speed implementation, new, semi-automatic approaches to information integration are required. With respect to leveraging information resident in a broad range of computer systems, Hammer and O'Brien (2004) divide the problem into knowledge extraction and knowledge composition. Knowledge extraction refers to identification of information in a given source, and involves technologies such as data reverse engineering (Aiken 1996), schema matching (Rahm and Bernstein 2001), program analysis (Paul and Prakash 1994), and data mining (Han and Kamber 2001). Each of these techniques attempt to automatically infer meaning about underlying data, and it is likely that a combination of techniques will form the basis of a toolkit for semi-automatic identification of knowledge. Knowledge composition refers to making use of the extracted data for use by other applications. Mediation is a common approach to knowledge composition (Wiederhold 1998). The services of mediators broadly include: assembly or fusion of information from distributed data sources; transformation of that data into another form (often increasing the information density by combining or summarizing raw data); and cleansing the data to improve its reliability or accuracy (e.g. resolving conflicting information, identifying and interpolating missing data, providing a confidence level of the mediated information). Each of these mediation services is necessary in the broad information environment of the digital infrastructure, which is likely to include multiple information sources that contain different levels of detail and often partially conflicting and/or incomplete information. The basic challenges of information integration are well understood, and various commercial knowledge extraction and composition tools exist. That said, the current state of the art largely consists of tools that help expert programmers speed their work on individual implementations. Semi-automatic tools remain very much in research status, limiting wide-scale deployment of an integrated digital infrastructure.

User Modeling

The concerns about human processing of information outlined by Brown and Duguid (2000), Norman (1998) and Schneiderman (2002) are addressed in the research literature on human-computer interaction (HCI) (Carroll 2003). The HCI field is large and contains much fundamental research. However, an HCI approach known as user modeling (Payne 2003) holds promise for rapid application to the digital infrastructure. Building from Rich's (1979) seminal work on stereotypes, user models are formal descriptions of users that can direct the efforts of both human software developers as well as automatic configuration of software. The promise of the user modeling approach is that it provides a formal description of the needs of a class of users (e.g. traffic operations personnel, police dispatchers, etc.). The description can then be used to customize for that class the presentation of the large number of data sources and applications that comprise the digital infrastructure. For example, problem solving environments (Rice and Boisvert 1996) combine rich analysis and visualization tools that could help managers make sense of the rich data in the digital infrastructure. Tools such as problem solving environments are generally limited to technical experts and are not easily configured to individual circumstances or made useful to non-expert users. User models hold the promise of directing such configuration for a broader range of users (O'Brien *et al.* 2003). User modeling approaches are still an area of research, but current practice based software development books detail approaches to describe user models (Cooper and Reimann 2003; Kuniavsky 2003). User models as active software agents are also being deployed to personalize software configuration (Strachan *et al.* 2000). A necessary step for deployment of the digital infrastructure is identification and development of user models for the wide range of users in the information environment in Figure 2.2. Hence, a user model for fire rescue personnel could aid assessment and presentation of relevant information from the broader digital infrastructure.

Information Security

A final note on technical issues for successful deployment of the digital infrastructure concerns information security. Information security as a field has recently evolved to include intelligence issues such as detection of terrorist activities (Chen *et al.* 2003). A deployed digital infrastructure of sensors could aid detection of terrorist actions, and it is likely that research in intelligence applications will help drive development of the information integration approaches detailed above. However, in terms of deploying the digital infrastructure, the more traditional definition of information security as the discipline of protecting the networked computing infrastructure is appropriate. Two issues must be addressed: (1) providing the correct access to individuals who are part of the information environment; and (2) securing the network from external eyes and assaults. The first issue may likely be overcome by a system of user models that describe information needs connected to a directory services architecture that assigns permissions for individual users (see Economist (2001) for discussion). The second issue is beyond the scope of this chapter and is the subject of intensive research and development activities worldwide. An excellent resource for information security issues is the SANS (SysAdmin, Audit, Network, Security) Institute (www.sans.org). See especially Lindner's (2001) primer on the issues of network security. Certainly, as the digital infrastructure is deployed, there will be an increasing need for securing the information contained within it. It does seem, however, that development of adequate security measures will keep pace with the needs of the digital infrastructure. Thoughtful deployment of the digital infrastructure should take into account security measures, but security concerns should not derail or unnecessarily delay deployment.

Conclusions – Towards a Fully Realized Digital Infrastructure

The digital infrastructure is multifaceted, offering considerable opportunities to radically increase the quality and timeliness of information about the physical infrastructure. Several imperatives – increasing demands on our physical infrastructure, visions for smart cities, security needs in a post Sepember 11 world – will drive the implementation of new technologies to leverage and secure our physical infrastructure. That said, it is unclear how quickly the digital infrastructure in its grandest forms will be realized. Certainly, much of the networking infrastructure is in place, and sensors for specific applications can be widely deployed quickly and at relatively low cost. However, sensors by themselves are not a panacea. For many applications, it will take considerable experience to learn how to reliably interpret and model sensor data. Broader uses for the information provided by sensors – uses that require integration among agencies in the information environment (Figure 2.2) – will likely take years to realize. Demonstration projects within municipalities that manually integrate information will be the first steps. Such demonstrations are necessary before widespread deployment as poor implementations may cause more problems than solutions by providing inaccurate information. As noted by William Mitchell (1996), the challenge is to create an environment that promotes "getting to the good bits" (Mitchell 1996: 162). Mitchell limits his argument to the interplay between architecture and the information network, but the implications are relevant to this discussion. The technical challenges detailed above prevent rapid adoption of the broader visions for the digital infrastructure. The technical challenges also suggest compromises and tradeoffs. It will take time, experimentation, and human judgment to guide deployment of the digital infrastructure.

Note

1 In Alachua County: Emergency Management, Facilities Management, Information and Telecommunications, Fire Rescue, Sheriff's Office, Public Works, and Environmental Protection. In Gainesville: Police, Fire Rescue, Public Works, and Gainesville Regional Utilities. At the University of Florida: Physical Plant Division, University Police Department, Environmental Health and Safety, and Office of Information Technology. Bell South also maintains the physical infrastructure for telecommunications in the region.

References

Aiken, P. (1996) *Data Reverse Engineering: Slaying the Legacy Dragon*, New York: McGraw-Hill.

Aktan, E., Catbas, F., Grimmelsman, K., Pervizpour, M., Curtis, J., Shen, K. and Qin, X. (2002) "Sensing, Communication, Computing and Information Systems for Infrastructure Health, Performance and Security Monitoring," White Paper, Drexel Intelligent Infrastructure Institute, Online. Available at: www.di3.drexel.edu/ (accessed November 5, 2003).

AWST (2003) Acoustic Wave Sensor Technologies, Online. Available at: www.army-technology.com/contractors/nbc/microsensor_systems/ (accessed November 3, 2003).

Brown, J.S. and Duguid, P. (2000) *The Social Life of Information*, Boston: HBS Press.

Campbell, G., Rogers, K. and Gilbert, J. (1995) "PIRAT – A system for quantitative sewer assessment," *International No-Dig '95 Conference*, Dresden, September 1995.

Carroll, J.M. (ed.) (2003) *HCI Models, Theories, and Frameworks: Toward a Multidisciplinary Science*, San Francisco, CA: Morgan Kaufmann.

Carter, W. and Shrestha, R. (1998) "Engineering applications of airborne scanning lasers: reports from the field," *Photogrammetric Engineering & Remote Sensing*, April, pp. 246–53.

Chen, H., Miranda, R., Zeng, D., Demchak, C., Schroeder, J. and Madhusdan, T. (eds) (2003) *Lecture Notes in Computer Science 2665: Intelligence and Security Informatics*, Berlin: Springer-Verlag.

CII (2003) "Emerging Construction Technologies," Construction Industry Institute, Online. Available at: www.new-technologies.org/ECT/ (accessed November 1, 2003).

Cooper, A. and Reimann, R. (2003) *About Face 2.0: The Essentials of Interaction Design*, Indianapolis, IN: Wiley.

Dry, C. (1994) "Matrix cracking repair and filling using active and passive modes for smart timed release of chemicals from fibers into cement matrices," *Journal of Smart Materials and Structures*, 13(2): 118–23.

Economist (2001) "List makers take control," *The Economist*, 360(8240), September 22, S27–9.

Elliott, J.F. (1999) "Continuous acoustic monitoring of bridges," *International Bridge Conference*, Pittsburgh, Pennsylvania IBC-99, p. 70.

Furuya, Y. (1995) "Design and fabrication of shape memory alloys," *Proceedings of ICCM-10*, B.C., Canada: Whistler, V-275–82.

Hammer, J. and O'Brien, W. (2004) "Enabling Supply-Chain Coordination: Leveraging Legacy Sources for Rich Decision Support," in Akçali, E., Geunes, J. Pardalos, P.M., Romeijn, H.E. and Shen, Z.J. (eds) *Applications of Supply Chain Management and E-Commerce Research in Industry*, Boston/Dordrecht/London: Kluwer Academic Publishers, pp. 1–47.

Han, J. and Kamber, M. (2001) *Data Mining: Concepts and Techniques*, San Francisco, CA: Morgan Kaufmann.

Hwang S., Trupp, T. and Liu, L. (2003) "Needs and trends of IT-based construction field data collection," *Fourth Joint International Symposium of Information Technology in Civil Engineering*, Nashville, TN, November 15–16, 2003.

Jaselskis, E.J. (2003) "RFID's Role in a Fully Integrated, Automated Project Process," Construction Research Congress, Honolulu, Hawaii, March 19–21, 2003.

Kuniavsky, M. (2003) *Observing the User Experience: A Practitioner's Guide to User Research*, San Francisco, CA: Morgan Kaufmann.

Kuntze, H.B., Schmidt, D., Haffner, H. and Loh, M. (1995) "KARO – A flexible robot for smart sensor-based sewer inspection," *International No-Dig '95 Conference*, Dresden, September 1995.

Lindner, C.E. (2001) "Information security primer," *Version 1.2e*, SANS Institute.

Livingston, R.A. (2003) "Nanotechnologies in infrastructure," Proceedings of two-day workshop, McLean, VA., Federal Highway Administration and Volpe Transportation Systems Center, April 2003.

Loftness, V., Hartkopf, V., Lee, S., Shankavaram, J. and Aziz, A. (2001) "Smart buildings, intelligent buildings," in Teicholz, E. (ed.) *Facility and Design Management Handbook*, New York: McGraw-Hill, 12.1–12.41.

McQueen, B. and McQueen, J. (1999) *Intelligent Transportation Systems Architectures*, Norwood, MA: Artech House.

Mitchell, W.J. (1996) *City of Bits: Space, Place, and the Infobahn*, Cambridge, MA: The MIT Press.

Norman, D. (1998) *The Invisible Computer: Why Good Products Can Fail, the Personal Computer Is So Complex, and Information Appliances Are the Solution*, Cambridge, MA: The MIT Press.

O'Brien, W., Issa, R. and Flood, I. (2003) "Moving from Information Tunnels to Configurable, User-Model Driven Environments: A Vision for Future Project Information

Technologies," *Fourth Joint International Symposium on IT in Civil Engineering*, Nashville, TN, November 15–16, 2003.

O'Brien, W.J., Issa, R.R., Hammer, J., Schmalz, M., Geunes, J. and Bai, S. (2002) "SEEK: Accomplishing enterprise information integration across heterogeneous sources," *ITcon – Electronic Journal of Information Technology in Construction – Special Edition on Knowledge Management*, 7: 101–24.

Paul, S. and Prakash, A. (1994) "A framework for source code search using program patterns," *Software Engineering*, 20(6): 463–75.

Payne, S.J. (2003) "User's mental models: the very ideas," in Carroll, J.M. (ed.) *HCI Models, Theories, and Frameworks: Towards a Multidisciplinary Science*, San Francisco, CA: Morgan Kaufmann, pp. 135–56.

Rabaey, J., Arens, E., Federspiel, C., Gadgil, A., Messerschmitt, D., Nazaroff, W., Pister, K., Oren, S. and Varaiya, P. (2001) "Smart energy distribution and consumption: information technology as an enabling force," Center for Information Technology Research in the Interest of Society (CITRIS), Berkeley, CA.

Rahm, E. and Bernstein, P.A. (2001) "A survey of approaches to automatic schema matching," *VLDB Journal: Very Large Data Bases*, 10(4): 334–50.

Rice, J.R. and Boisvert, R.F. (1996) "From scientific libraries to problem-solving environments," *IEEE Computational Science & Engineering*, 3: 44–53.

Rich, E. (1979) "User modeling via stereotypes," *Cognitive Science*, 3: 329–54.

Roush, W. (2001) "Networking the infrastructure: the technology of redundancy could lead to smarter cities, designed to alert us of danger," *Technology Review*, December, pp. 38–42.

Schneiderman, B. (2002) *Leonardo's Laptop: Human Needs and the New Computing Technologies*, Cambridge, MA: The MIT Press.

Srinivasan, A.V. and McFarland, D.M. (2000) *Smart Structures: Analysis and Design*, New York: Cambridge University Press.

Strachan, L., Anderson, J. and Sneesby, M. (2000) "Minimalist user modeling in a complex commercial software system," *User Modeling and User-Adapted Interaction*, 10: 109–45.

US Environmental Protection Agency (2002) *The Clean Water and Drinking Water Infrastructure Gap Analysis*, EPA-816-R-02-020, September.

Wiederhold, G. (1998) "Weaving data into information," *Database Programming and Design*, 11(9).

Willke, T. (1998) "Five technologies expected to change the pipe line industry," *Pipe Line & Gas Industry*, January, 81(1).

CHAPTER THREE

Management Challenges

Thomas Horan and Roy Sparrow

Introduction: Digital Romans

In AD 84, Sextus Julius Frontinus, Roman water administrator proclaimed, "I also lay aside all ideas of any new works . . . the invention of which long-ago reached its limit, and in which I see no hope for further improvement" (Frontinus AD 84; Evans 2000). Two thousand years of public works projects have proven Frontinus wrong, as we have devised civil and environmental infrastructures that exceed the scale, complexity, and expectations of the Roman waterworks.

Still, even in Roman times, the technical and management challenges were considerable. The construction of the Roman water infrastructure involved eleven major aqueduct routes and took over 500 years to accomplish (from 312 BC to AD 226). Importantly, Hall (1998) observes that the aqueducts were less a matter of technological marvel and more a matter of competent management: "the Romans achieved no major technological breakthrough of their own. Rather, they perfected and systematized the techniques, acquiring an extraordinary everyday competence in civil engineering" (Hall 1998: 637), all of which, Hall notes, was supported by "a complex system of administration. An important part of which was concerned with the nuts and bolts of everyday service provision" (Hall 1998: 653).

If the Romans were to reconstruct the aqueducts using today's vernacular, they would analyze the enterprise as a "complex system," utilizing "high-level" modeling of system interactions. The detailed hand-etched renderings would be replaced by an elaborate IT infrastructure providing CAD, GIS, and other information tools and products. There would be multiple organizations involved, some responsible for the planning, many responsible for operations, and a bureaucracy that would rival if not surpass the Roman army!

What does it mean to manage such infrastructures under contemporary circumstances? As a point of departure, it is important to recognize that history serves as more than illustration when it comes to understanding contemporary infrastructure. Indeed, the accidental history of social, economic, and technological factors combine to create the circumstances under which infrastructures grow and evolve. This view, as espoused by Arthur (1994), both helps understand past events as well as serves as a cautious guide to the future:

> Thus, it is highly likely that the system of cities we have inherited is only partly the result of industries' geographical needs, raw material locations, the presence of natural harbors and transportation costs. It is also the result of where immigrants with certain skills landed, where early settlers met to market foods, where wagon trains stopped for the night, where banking services happened to be set up, and where politics dictated that canals and railroads be built. We therefore cannot explain the observed pattern of cities by economic determinism alone without reference to chance events, coincidences and circumstances in the past; and without knowledge of chance events, coincidences, and circumstances yet to come, we cannot predict with accuracy the shape of urban systems in the future (Arthur 1994: 109).

As just one example of the power of circumstances, consider these recent challenges to infrastructure systems in North America (Connery and Clevenger 2002):

- regulatory shifts toward greater privatization and decentralization
- dramatic swings in local and global economic conditions that alter the mix of capital and operations activities
- changes in the financing of both capital investment and operations toward increased state-local governmental and private sources
- shift from an emphasis on new infrastructure toward rebuilding and upgrading existing infrastructure
- growing performance expectations by users and customers
- growing awareness of the interdependence of infrastructure systems with each other and with the natural and built environments
- transformative change in the means by which technology is integrated into systems design and deployment
- pronounced security concerns in light of recent waves of terrorist attacks.

Such contextual developments require of organizations and their leaders far greater capacity for vision, adaptability, innovation, and collaboration than in previous decades. Not surprisingly, infrastructure organizations struggle to improve infrastructure services at a time when the context, demands, and technologies are rapidly changing. To some observers, this speed of change, together with the unpredictability of chance occurrences, produces a growing risk that we are creating systems that we cannot control (Winner 1977). The challenge to managers is serious indeed.

This chapter confronts this challenge by first outlining key elements for managing digital infrastructures (drawing generally on complexity theory) and then applying these concepts to major civil and environmental systems. It then addresses the training challenges relative to infrastructure managers and closes on a broad notion of design as a means to bring innovation to both the practice and pedagogy of digital infrastructure management.

Overview of Complex Digital Infrastructures

A defining characteristic of the systems described throughout this volume is their complexity. In this sense, complexity refers to the fact that their character is shaped by myriad contextual and endogenous factors that are dynamic and emergent and their evolution is adaptive. Such a perspective draws heavily on approaches such as by Dooley (1996), who offers the following definition of a complex adaptive system (CAS): "A CAS behaves/evolves according to three key principles: order is emergent as opposed to predetermined, the system's history is irreversible, and the system's future is often unpredictable." Others have elaborated on these principles. Anderson (1999), in particular, provides a seminal analysis of implications of CAS on organizational functioning. He outlines several dynamics that should guide an organization's functioning; these include: (1) dis-equilibrium is the norm, (2) external agents drive the process, (3) complex processes are affected by small change, (4) complex processes resist reductionism, (5) patterns emerge from simple rules, and (6) systems exhibit self organizing characteristics. As Anderson observes, these principles have deep roots in organizational theory (including open systems theory and earlier views of cybernetic systems). Complexity theory extends these works to emphasize the decentralized, adaptive, context-rich behavior of organizations (Anderson 1999). Moreover, as will be taken up later in the chapter, it suggests an approach for the broad *design* of organizational strategies and management practices.

The key premise of organizational complexity theory is that managers and organizations that understand these dynamics will be better equipped to address what can appear to be an unwieldy set of ever moving factors. In this chapter, the management domain under consideration is digital infrastructures. Figure 3.1 provides a conceptual model for this domain. As illustrated in the figure, there are five dimensions that comprise digital infrastructures: infrastructure users, infrastructure elements, information systems, managing organizations, and environmental conditions. These are each discussed in terms of the management challenges they pose for infrastructure managers.

Infrastructure Users

Any system discussion should begin with a consideration of the systems users. While private businesses have developed rather sophisticated means to attend to

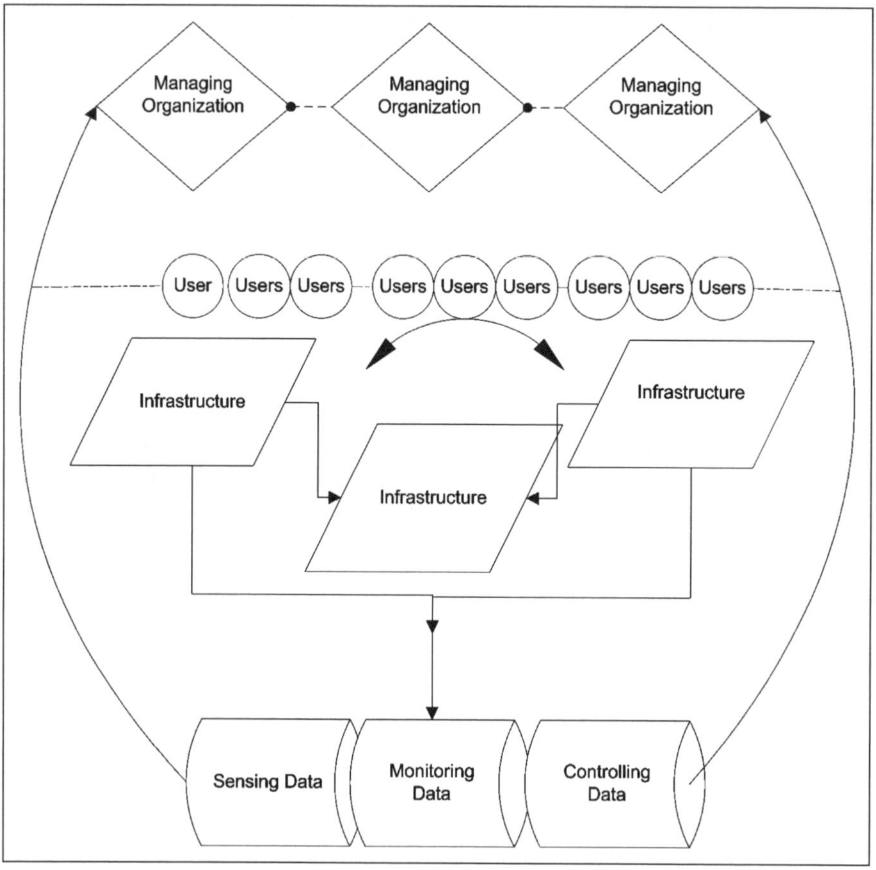

Surrounding Social, Economic, Policy, and Physical Environment

3.1 *Conceptual Model for Managing Digital Infrastructures.*

customer interests and desires, from an infrastructure management perspective, users have typically been considered from the abstract perspective of "demand." For example, there is travel demand on transportation systems or electrical demand during peak periods on energy systems. However, users need to be viewed as a highly differentiated set of interests if infrastructure systems are to be managed efficiently and effectively (Horan 2003). In electricity, for example, there is a range of residential and institutional customers with quite different needs for surge capacity, service reliability, etc. From the perspective of complex systems, users play essential roles as agents that drive system change and performance. For example, the recent behavior of mobile phone users in rapidly adopting and using e-911 for roadway emergencies has profoundly affected the nature and performance of EMS systems (Horan and Schooley 2004).

Infrastructure Elements

At the heart of a digital infrastructure system is a physical system to be managed. This volume features several – transportation, electricity, water, telecommunications. From a management perspective, their complexity suggests a number of important attributes – each system's behavior cannot be fully predicted, each system's subsystems are tightly linked together and so there can be cascading impacts, each system is dynamic and will never reach a fixed-point or a cyclical equilibrium, and a small shift in a parameter of each system could send that entire system to a different equilibrium. Amin (2000) analyzes the challenges poised by interconnected infrastructures in terms of safety and reliability. He finds that, "No single entity has complete control of these multi-scale, distributed, highly interactive networks, or the ability to evaluate, monitor, and manage in real time.... The networks' interconnected nature makes them vulnerable to cascading failures with widespread consequences (Amin 2000: 263–286). Consequently, it is not surprising that physical infrastructures often provide "textbook" examples of massive system failures, such as electricity blackouts or telecommunications gridlocks.

Information Systems

The third critical feature for digital infrastructure management is the information system that provides data on the infrastructure's performance. As denoted in Figure 3.1, this data can cover the gamut of sensing, monitoring and controlling functions. While in the ideal these data would be available through a centralized database, more typically these information sources are highly heterogeneous, distributed and of varying quality. For example, in surface transportation, modal information (e.g. transit, highway, freight) is kept separately by different agencies and the underlying system performance data is often truncated for use by municipalities and other users. From a management perspective, these information systems require a range of technical competencies by infrastructure professionals, in terms of IT planning, deployment and maintenance. Moreover, these management challenges are compounded by the vulnerability of infrastructure systems to IT failure. As Park and Chandramohan (2004: 1) observe:

> In some cases, lives and livelihoods depend on the continuation of IT systems that monitor and maintain critical activities such as traffic control, telecommunication, banking and power management. As information systems become more complex and we rely more on their correct operation, the need for survivable systems becomes ever more pressing. Furthermore, the interdependence of these systems complicates the survivability picture.

For this reason, considerable effort needs to be devoted to understanding how such information systems can be used to avert rather than contribute to system vulnerability and failure.

Managing Organizations

Users are not the only critical agents in the system. Organizations act as agents charged with managing the operational and user needs of the system. However, a common feature of digital infrastructures is the cross agency linkages. Rarely does a single organization manage all infrastructure elements. These partnerships can vary by geography (e.g. agency partnerships across adjoining regions), by function (e.g. different organizations responsible for planning, building, maintaining and operating infrastructures) or scale (e.g. partnerships between local systems and regional agencies within a geographic area). While many infrastructure organizations already have a strong capital-building component, from a complexity management perspective, the new priority should be on building and maintaining their inter-organizational services and information networks. It is worth underscoring that these inter-organizational networks need to work across a highly fragmented set of organizations. Moreover, these public and private organizations often have differing information systems and data use priorities. Fountain (2001: 200) comments on the formidable challenge of managing inter-organizational IT networks, "Public managers must learn to ... master the network game if they are to establish productive and useful partnerships in a time of scarce and diminishing government resources and rapidly changing environments . . . the Internet, in its role as catalyst, infrastructure, and tool, adds another layer of complexity."

Infrastructure Environment

Surrounding these four elements is a natural, social, economic, policy and technological contextual environment. The introduction to this chapter noted the role that history plays in creating conditions for infrastructure development. The concept of organizational environment has been recognized for a long time as a subtle, but important driver and shaper of management (Scott 1981; Pfeffer and Salancik 1978; Katz and Kahn 1978). Sussman and Dodder (2002) nicely summarize the critical role of the contemporaneous environment:

> Nested complexity suggests that there is a physical system ... embedded within a much messier sphere representing the policy system. This sphere represents the organizational and institutional framework of policy-makers, firms, stakeholders, non-governmental organizations, and others that together comprise the broad policy system.

Other crosscutting chapters (e.g. policy) in this volume address various aspects of the contextual environment, so it is not addressed in detail here. However, this is not to understate its criticality. Indeed, environmental assaults (in the sense of creating disequilibrium) are often the driving factors that both explain an infrastructure's dynamic functioning and condition the nature of operational and strategic approaches to improving such an infrastructure's performance.

While it may be a useful exercise to speak to these five distinct components, the power (and the problems) of digital infrastructure largely arise because of the interactions of these elements. For example, one has to appreciate the interactions between users, organizations, information systems and environment in order to understand how "perfect storm" events such as cascading power failures, wastewater overflows, or telecommunication jams can occur – sometimes all at once.

Real-time Digital Infrastructure Management

"Normal" Conditions

The following two examples illustrate these five elements, as they are operative in the field today. The first case exemplifies the role that users can play in driving new infrastructure services once they are provided with innovative information and service options. It also illustrates how realizing such an opportunity requires the dynamic interaction of users, managing organizations, and information systems.

The second case demonstrates the role that managing organizations can take to ensure infrastructure performance in times of crises. As elegantly demonstrated in Perrow's *Normal Accidents* (1999), crises events provide a unique lens to understand how organizations and systems have structured themselves to facilitate and respond to incidents and accidents, as well as provide the means for rapid response and resolution. Moreover, from an infrastructure performance perspective, dealing with crises or extreme events can be a pivotal test of the overall system management capability

Case Example 1: User-driven Service Improvement

Gothenburg, Sweden, is home to 500,000 residents and some of the most innovative real-time information and traffic management systems in the world. The city's Flexline service was created to provide minibus service for the elderly and mobility-challenged. A fleet of accessible minibuses and an advanced IT system make this high-performance, cost-effective transit service possible.

For years prior to Flexline's development the city had offered its mobility-challenged customers a demand-responsive taxi service. And, as in most other urban regions in the economically developed world, dedicated fleets provided social service transport. The taxi option had proven very expensive. Flexline's innovation (begun in 1996) was to use digital technology to create a hybrid service with features of both demand-responsive and fixed routes. "The service is fully flexible within established end-points. Rides are booked in advance through a computer that determines the best routes and approximate travel times" (Marie 2001: 5).

The entire network of public transport in Gothenburg is controlled by a unique traffic management and information system that oversees the buses and trains for the region around Gothenburg. The brain of this mobility management operation is a software package that virtually eliminates manual route planning and dispatching for most services, including Flexline. Customers eligible for Flexline service, for example, simply phone the computer and punch in their personal identification number. The computer program maintains a running total of the number of travelers already booked for a given trip and calculates whether one additional stop will cause the bus to be later than its allotted time. If the request meets time and capacity standards, the computer redesigns the planned route and promptly calculates the time that the rerouted vehicle will arrive at the designated meeting point. The traveler receives a phone message from the automated system 15 minutes before the agreed pick up time.

The return trip is simpler yet. At the destination, such as a shopping mall, the user swipes an identity card at a designated location. This initiates communication with the central computer system that recognizes the individual end location. The automated dispatcher assigns the rider to the next available trip home. An automated voice then communicates the time that the next available bus will arrive at the customer's location. Round-trip travel service for a wide variety of transportation-challenged and public transportation-dependent users has been provided without the cost of telephone operators or vehicle dispatchers.

Significantly, the Flexline service is but one of many modally integrated and inter-modally connected transportation services developed in Gothenburg and supported by its unique traffic management and information system. In fact, Flexline is merely the latest application of Gothenburg's innovative digital routing/dispatching system that manages all services other than the traditional route system. According to Coogan (2000: 7), the centerpiece of this mobility management operation is software that,

> . . . virtually eliminates the separate tasks of manual route planning and dispatching. The client calls in to a telephone operator, and gives an identification number, which in effect, establishes his/her right to receive this kind of service. The telephone operator enters the origin, destination, and desired pick up time, and in a fraction of a second the telephone operator's screen assigns the trip to a service providing contractor, and informs the user of the pick up time. At this point the algorithm may or may not have determined the total number of riders who will share this trip. To the extent possible, riders for whatever purpose are grouped together to make the trip more cost effective. A higher volume assembled will be assigned to a van, while the smaller loads are assigned to a taxi. Over 6,000 trips a day are dispatched for social service agencies ranging from hospitals to schools to elder care.

Every three years the managing agency issues a tender seeking bids for the provision of taxi and group services. This transportation system operates without governmentally owned vehicles and the drivers are likewise private. Yet the government takes responsibility for the provision of extensive, highly integrated transportation, made possible by digital innovation.

Case Example 2: Interorganizational Leadership in the Face of Crises

In October 1997, a tanker explosion in Yonkers, New York in a viaduct beneath the New York State Thruway closed that vital north-south artery and threatened chaos on linked interstate highways and bridges throughout the northeastern US That this did not occur is, in significant part, attributable to the unseen hand of TRANSCOM, a multi-state, multi-jurisdictional coordinating agency that possessed an IT network that allowed the agency (through its inter-organizational network) to send instant warning messages along the highways to points as far away as Maryland and Massachusetts so that drivers could plan alternative routes. That the repair was completed in a matter of weeks, not months, was also partly attributable to TRANSCOM's ability to immediately get agencies to reschedule planned repairs on arteries in the vicinity of the explosion, thus permitting repair of the vital Thruway to take priority. This was feasible because TRANSCOM works to streamline multi-jurisdictional operations not only during emergencies, but also on an ongoing basis.

Since its beginning in 1986, TRANSCOM (Transportation Operations Coordinating Committee) has developed an IT and operational capability to acquire and use real-time information to improve the flow of traffic throughout the northeastern US and has now evolved into a vital component of traffic management for the tri-state region around New York City. In essence its mission is to make workable, indeed to enhance, the missions of its 18 transportation and public safety agencies by using IT to increase the reliability and speed of information in the possession of the individual member agencies. It has become so adept at this work that it has become a nationally recognized test bed for the implementation of intelligent transportation technologies in a multi-jurisdictional environment (TRANSCOM 2003).

Digital technology has made possible the expansion and enhancement of the operational capabilities of individual infrastructure organizations. More importantly, perhaps, it has made possible the linking of operational capabilities of these organizations, permitting far greater effectiveness and efficiency. While most agencies remain committed to more traditional bureaucratic approaches to goal setting and service production, some organizations (and collaboratives) have begun to exploit the advantages of digital technology at the operational level (Gifford and Pelletiere 2002). In the case of TRANSCOM it has enabled the evolution of a dynamic new regional transportation entity that was inconceivable prior to advanced computer technologies. The success of such organizations is critically dependent on new organizational structures and IT capabilities that will allow for an overall systems view of performance and the tools and relationships to make inter-organizational cooperation happen. These issues are taken up next.

Using End-to-End Performance to Drive Extraordinary Management

Strategy Under Uncertainty

Infrastructure managers must often adopt long timelines when considering new investment and deployment. Transportation, water, power and the like are capital intensive systems, investments in which need to be planned over a long time horizon. Moreover, the various regulatory, political and social-environmental factors all create pressure to adopt a "long-haul" perspective.

These conditions call for a strategic orientation under conditions of high uncertainty. The management requirement for dealing with this uncertainty is what Stacey (1993) refers to as "extraordinary management." In comparison to "ordinary management," which takes a near-term, operational, and logical analytical style, extraordinary management requires the development and use of expertise, tacit knowledge and creative problem-solving through communities of practice that self-organize around defining issues and needs (Stacey 1993; Wenger *et al.* 2002).

The third case of wireless emergency medical (e.g. 911) services provides an example of an infrastructure service that has typical digital infrastructure features: it involves multiple infrastructure and service providers, is customer centric, is highly dependent on technology, is the result of historical chance and certain dynamic developments in the marketplace, and is guided but not fully predicated on public policy. Moreover, this case provides an illustration of the management challenges when moving from an operational focus to a strategic focus and the role that digital infrastructures can play in aiding this transition.

Case Example 3: Information to Guide End-to-End Strategy Development

When the cell phone was first introduced in 1983, it was considered (and marketed) primarily as a business tool. Even with demand outstripping expectations, rigorous estimates were done that suggested a saturation point would be achieved in relatively short order. Appearing in a 1995 *Communications of the ACM*, Wang and Kettinger predicted cellular phones would reach saturation at 80 million subscribers.[1] There are currently about 150 million US subscribers and that number is growing (CTIA 2003). Beyond sheer growth in the use of wireless service, there has been growth in the types of use. While articles like Wang and Kettinger focused on business use, the subsequent and evolving diffusion has come to entail a variety of uses. Among these developments has been the rise of cellular phone as the "safety net" for travelers: 911 calls from cell phones have risen exponentially from under 10,000 per day in 1989 to approximately 200,000 per day in 2003.

The institutional infrastructure to deal with this rapid rise in emergency cell phone use is barely able to accommodate these demands. Even in the post-September 11 environment (and related emphasis on first responder systems), the public service answering system is struggling to upgrade (NENA 2003). For example, whereas 97 percent of local (911) public service answering points (known as PSAPS) have the ability to determine the location of wireline 911 calls, only four percent have the ability to determine the location of wireless calls.

As noted in a recent analysis of Emergency Medical Services (EMS) (Horan and Schooley 2004), the mobile EMS process represents a complex adaptive system where technology, organizations, and policy interact to create a dynamic system where multiple agents operate to deliver a customer service under critical time pressures. Drawing on Stacey's distinction between ordinary and extraordinary management, the ordinary management task is highly operational. This task is to deliver "end-to-end" EMS service. At one "end," once a wireless 911 call comes in, there is a time critical need to identify the location of the emergency and the requirements in terms of service. Next, there is the need to deliver that service rapidly through emergency providers, such as ambulance services. Finally, at the other "end," there is the emergency-medicine related need to provide critical services (in the ambulance or emergency room) in the quickest time possible. Each of these service functions depends upon IT (and related communications) networks.

While these ordinary tasks are subject to structured rules and systems for "end-to-end" delivery, the larger "extraordinary task" is to look forward strategically to how the entire system might function in the future. This more strategic view calls for infrastructure service providers to step out of their "silos" and examine the information flows between services. User demands, technology systems, and policies all intersect to create an evolving set of service needs that should be addressed simultaneously. For example, new locational information in the cell phone requires an upgrade in the telecommunications system, which in turn has implications on local funding policy and procedures.

IT systems can facilitate "end to end" performance improvements by the chain of EMS delivery organizations involved in the EMS process; these organizations range from the state patrol office that may first receive the emergency call to the ambulance service that delivers the patient to the hospital. Using insights gleaned from information systems on "end-to-end" performance, this network of providers has the potential to lead a more strategic view of how to organize emergency detection and response (ComCare 2003). For example, in northwestern Minnesota, transportation, state patrol, and emergency providers are collaborating on an integrated Transportation Operations Communications and Control (TOCC) system that will encourage collaborative development of service information systems. In Shenandoah Valley, Virginia, the health care providers are taking a lead role in E-Safety Network Initiative, a program designed to create a real-time information system that all local response agencies can use to provide and track service through the entire "end-to-end" process (thereby replacing voice communications as the principle means of service coordination).

Initiatives such as the TOCC program in Minnesota and the E-Safety Initiative in Virginia challenge organizations to evolve organizational and performance patterns through innovative IT systems. Whether such organizations can actually evolve depends on the nature, style and success of organizational learning that occurs. The final section of the chapter turns to this learning challenge.

Training Digital Infrastructure Managers

New Core Competencies

As suggested by this chapter's conceptual model and case examples, achieving the promise of digital technology will require more of infrastructure organizations than just good physical systems planning and design. In recent years, a number of conferences have considered the matter of core competencies for infrastructure professionals (ICIS 2000; Sparrow 2001; Yao and Roesset 2001). A related set of research has focused on the educational and training needs of professionals and managers in areas of infrastructure heavily impacted by IT (Volpe National Transportation Systems Center 1999). A common finding of these works is that a broader set of competencies is needed. Drawing on these works, the following summarizes how these key competencies apply to managers of digital infrastructure systems.

Technical competency remains important. But as revealed in the digital infrastructure conceptual model, this technical competency needs to include infrastructure systems *and* the IT systems that support these infrastructures. In terms of the former, the traditional domain knowledge of civil and environmental systems is still critical but needs to be augmented with a competency in planning, managing and assessing dynamic individual and converged infrastructures. In terms of the latter, managers need to understand the role that information systems play in planning, monitoring, and controlling infrastructures, including during crisis conditions. This competency would include an understanding of the underlying sensing, monitoring and control technologies, as well as data integration and analytical systems to facilitate performance monitoring and enhancement.

However, a clear finding from a conference on infrastructure educational needs was that technical systems competency is not sufficient (ICIS 2000). As highlighted in the case studies, competencies are required in understanding user needs and organizational capabilities for addressing these needs. While the software industry has begun to understand the importance of user satisfaction with online and other IT services (Ivory and Hearst 2001), such a perspective is relatively new to infrastructure professionals (Horan 2003). This competency includes the ability to assess and plan for diverse user needs across digital infrastructures as well as the ability to operationally and strategically plan organizational adoption of IT systems to support civil and environmental infrastructure services. Moreover, nascent work on e-governmental networks has emphasized the importance of "inter-organizational systems" to ensuring the seamless performance of governmental agencies (Black *et al.* 2002). For the infrastructure manager, this suggests the need for communications skills in creating and maintaining inter-organizational networks, such as those described in the TRANSCOM case study.

Beyond the user and organizational systems lies the domain environment. Digital infrastructure managers must have a competency in recognizing the implications of the natural, social, environmental and policy milieu on system performance (Sparrow 2001). This includes, for example, a broad knowledge of history (as related to infrastructure and communities), sociology (as relates to users and communities), public policy (as relates to infrastructure and IT policies) and environment (as relates to environmental trends and conditions). Perhaps most importantly, this includes the ability to integrate knowledge across such a wide range of subjects. Because IT pervades the infrastructure system, this integration task must now include an assessment of how IT related infrastructure services will work within the deployment context; that is, how the nature, style and range of information services for the infrastructure should be tailored to meet local user and community demands.

Learning Approaches

The successful infrastructure manager must be engaged in constant learning, both collaboratively in the work environment and through formal training programs. Over the past decade researchers in the field of continuing professional and executive education have documented both the need for expanding the managerial mindset and practice-based approaches to accomplishing lasting results. "Corporate universities" that directly serve corporations facing rapidly changing competitive pressures have pioneered integrative and action-oriented methods to keep managers ahead of the curve. Robert Fulmer, a leading authority on executive education, notes that "[m]ost corporate universities share a major focus on building competencies and skills that are strategically aligned to meet the requirements of employees and the company" (Fulmer 1997: 62).

For infrastructure managers, this means creating a "learning organization" that understands the role of IT in creating decision-oriented information. In his seminal work on learning organizations, Senge (1990) contended that in organizational systems, thinking and learning was poorly understood and little practiced. In the decade since Senge's work appeared, research has only reinforced his prescription for engaging managers in systems thinking and learning (Flood 1999). His "component technologies" are as important for the world of digital infrastructure as for the corporate environment he addressed – pedagogy is more likely to achieve results to the extent that it is systems-oriented, interdisciplinary, and experienced-based. The implications of these dimensions for learning relating to digital infrastructures are myriad:

- Systems-oriented. Infrastructure professionals need to learn to pay attention to the whole being created even as they contribute their technical expertise to a particular component (Flood 1999). As Anderson (1999) notes, the advent of advanced IT systems now provides managers with the capability to obtain real-time information about how the entire system is performing. The learning challenge is becoming how to detect patterns within an ever-increasing amount of information. For the infrastructure professional, this requires competency in a range of analytical methods, from qualitative contextual mapping to quantitative analysis of performance information. It also requires an ability to perceive the information needs of a broad spectrum of stakeholders associated with different aspects of infrastructure systems.

- Interdisciplinary. In practice, most professional and managerial work involving infrastructure systems is interdisciplinary. The more professionals and managers can see the larger systems (temporal and spatial) toward which they are contributing, the more they can appreciate the inherently interdisciplinary, interfunctional, and interorganizational nature of their enterprise (Lattuca 2001). The learning challenge is to develop an efficient means to interact with concepts and perspectives from other disciplines and professions. In terms of digital infrastructures, a key interdisciplinary connection needs to be built between civil and environmental systems and IT management perspectives. While the former has tended to be "macro" in its orientation (e.g. environmental systems data) and the latter "micro" in its orientation (e.g. corporate enterprise systems), the need exists to devise means by which organizations can evolve "meso-level" strategies by using broad-based system data to create effective organizational strategies.

- Experience-based. Experience can be an effective teacher and presently many organizations are satisfied to let new infrastructure professionals and managers learn on the job what has not been acquired in school. However, experience needs to interact with knowledge from experimentation and research; this is particularly true given the fast changing and technical basis of IT systems. The task for trainers and educators is to design a set of learning goals built around a whole facility or project and challenge students to figure out the best and appropriate role for IT systems to produce information that decision-makers need to effectively manage the subject infrastructure (including its interactions with other infrastructures). Progress in this direction means that faculty need to discover, test, improve and deliver methods of situation-based digital infrastructure education that few have much experience with (Klein 1994).

By following a learning approach that is system-oriented, interdisciplinary, and experience-based, infrastructure managers will be better able to consider digital infrastructures within a broader context of performance. Moreover, such an approach will also give rise to demand for new skill sets at the interface between infrastructure management and IT. This would include, for example, demand for skills in IT project management, managing vendor and (out)sourcing relations, and devising multi-leveled management information systems (MIS) for infrastructures (Johnson 1999; Volpe National Transportation Systems Center 1999). While some of these skills overlap with general infrastructure management competencies, the IT dimension tends to be quite dynamic and therefore requires deliberate attention in both formal and informal training environments. As discussed in the final section, this has significant implications for the entire physical infrastructure profession.

Professional Implications
Professions such as civil and systems engineering and management evolved from reliance on approaches that solved complex problems using ingenuity and technology to one that teaches specialized methodologies for solving problems. A return to at least the philosophy of the earlier approach may be in order. The current output of educational programs is a properly socialized, technically competent, standardized professional, one who can be plugged into the on-going operational environment of any firm or organization at work on infrastructure. Even in the more loosely structured professions such as management, planning and policy analysis, there exists a similar pull toward specialization and standardization of product. Yet, the advent of digitally infused and interdependent infrastructures requires both a technically-grounded systems knowledge, as well as a breadth of knowledge across domain areas – this is the new infrastructure professional (Sussman 1995). This new infrastructure professional will operate amidst global trends, trends that need to be understood to successfully devise digital infrastructure systems. For example, in *Civil Engineering Practice for the Twenty-first Century*, Grigg *et al.* (2001) highlight eight such trends that substantially affect the civil engineering profession: globalization, population-development, *technology and knowledge*, governmental changes, social shifts, business dynamics, and new work forms.

Training this new profession requires a rethinking of professional education and the profession itself. Several studies have recently examined the challenges faced by the institutions that train infrastructure professionals (most commonly engineering schools and programs) in the United States. For example, the expert panel convened for *Engineering Education*, published by the National Research Council (1995), agreed that several skill areas deserve priority attention: development of communication and social skills; the need to integrate social and economic studies and liberal arts into the engineering curriculum; and the need to prepare students for career-long learning. The present volume on digital infrastructures aims to update this list by highlighting the important role that IT plays in achieving and maintaining high performance and critical infrastructure systems.

IT is changing the nature and context of work in infrastructure systems and educational programs preparing civil infrastructure professionals and managers must change to better prepare the workforce to meet these challenges. Reforms are needed in both the substance of the knowledge (e.g. core competencies) as well as the style by which this knowledge is obtained (e.g. learning approaches). A number of universities have initiated innovative programs with elements of interdisciplinarity and experienced-based learning (ICIS 2000). They are a start. But there is still a long way to go.

Conclusion

From Elements to Principles

Managing complex infrastructure systems enabled by IT presents daunting challenges. Indeed, the notion of managing complex adaptive systems (CAS) stretches traditional concepts of what "managing" and "management" mean. In the complex world of infrastructure systems there will always be much that is unknowable, unpredictable, and ultimately unmanageable, in the traditional sense of beyond the real control of the individual (or individuals) with nominal authority. "The most we can do," suggests Robert Flood, "is to manage what is local, whilst appreciating the incomprehensibility of global complexity. Managing what is local entails continually considering outcomes that extend over a small number of interrelationships, very few stages of emergence, over only short periods of time into the future" (Flood 1999: 90). Complexity requires systems and systems of organizations that are capable of learning (hence the term adaptive) (Senge 1990).

If infrastructure managers are to take advantage of the promise of digital technologies, there is no getting around the need to acknowledge if not embrace the elements and dynamics associated with complex adaptive systems (CAS) as applied to civil and environmental infrastructures. Reflecting once again on the conceptual model that underlies this chapter, management principles that flow from this orientation include:

- *Users:* recognizing the critical role of the customer in the service delivery process and organizing the service, including IT, to take advantage of it

- *Infrastructure Elements:* considering infrastructure elements as dynamic and evolving artifacts and moving away from a capital management orientation to a performance management orientation that utilizes IT to achieve high performance

- *Information Systems:* recognizing the critical role IT plays in the real-time management of systems and moving toward transparent data on end-to-end performance that can be used to manage infrastructures in normal and crises conditions

- *Managing Organizations:* accepting that positive organizational outcomes and good performance are fundamentally about strategically managing well-trained people, dynamic systems and IT networks in the face of uncertainty

- *Environment:* maintaining a broad awareness of social, technological, environmental and policy developments, as well as anticipating how IT systems can contribute to the prevention, anticipation and resolution of context-driven disruptive events.

As these principles suggest (and as discussed above), training and educational endeavors need to incorporate both broad review of IT developments as well as specific learning elements related to IT use in civil and environmental systems. These elements should provide for critical skills to manage IT planning, deployment and maintenance including interfaces with users, managers and other infrastructure organizations. Finally, the fast-changing nature of IT serves to underscore the importance of life-long learning to individual and professional performance.

Toward Design 2.0

Hebert Simon (1996) viewed the management challenge of complex systems as a problem of design. He wrote, "Engineering, medicine, business, architecture, and painting are concerned not with the necessary but with the contingent – not how things are but how they might be – in short, with design" (Simon 1996: xii). More recently Walls *et al.* (1992) and Markus *et al.* (2002) have extended this design thinking to outline a concept and method for designing information systems in service to organizations charged with managing emergent processes. An important observation made by Walls and colleagues is that design (and design science) is both a "noun" and a "verb," as it pertains to both a process of design and management as well as the resultant artifact.

In the case of Markus *et al.* (2002), the authors suggest that emergent knowledge processes are critical when decision-making is either semi-structured or unstructured. In this sense, the design process is interwoven with the creation of the artifact. Boland (2002) dis-assembles this recursive relationship into variations of three management functions identified by Simon: intelligence choice-design. Moreover, Boland highlights the *design*-intelligence-choice sequence (as compared, for example, to intelligence-choice-*design* sequence) as that sequence most in tune with evolutionary sense making in organizations. That is, one begins with the first order design issue and then gathers intelligence so as to inform the strategic choice of organizational evolution. In the case of infrastructure management, there is already a legacy of *design* in place. The strategic *design* (i.e. "design 2.0") issue is how to evolve this infrastructure in the face of uncertainty, using digital infrastructure to enhance the *intelligence* via information flows, so as to enable informed management *choice* of organizational evolution and learning steps.

The current notion of design in civil engineering has a strong pen and pencil connotation. But, a broader notion of design is needed. This notion suggests a digital infrastructure design process that incorporates elements of users, systems, organizations, as well the social-political context. Managers of these systems need to develop a holistic view of entire systems performance, have skills in inter-organizational and dynamic management and be trained in a setting that provides context-rich experience in complex system design, management, and operations. This is a much larger notion of design, but one that befits the nature of the times we live in.

Acknowledgments

The chapter benefited from background research undertaken by several student research assistants, including Ben Schooley, Albert Sargent, and Peter Bernardy. Several of the ideas expressed in the chapter were presented in preliminary form at the University of Minnesota, Humphrey Institute's Workshop on Complexity and Transportation Management, ICIS's Workshop on Bringing Information Technology to Infrastructure, and ICIS's Workshop on Civil Infrastructure Systems Education. The authors gratefully acknowledge the value of these forums in crystallizing the concepts that appear in this chapter.

Note

1 In fairness to Wang and Kettinger, they did conclude their article on the following note: "While it is of greater planning value to make projections concerning newer phenomenon earlier in their adoption curve, one should take care in interpreting forecasted numbers generated by our model, especially in long range projections." (Wang and Kettinger 1995: 122).

References

Amin, M. (2000) "National Infrastructures as Complex Interactive Networks," in Samad, T. and Weyrauch, J. (eds) *Automation, Control, and Complexity: An Integrated Approach*, New York: John Wiley and Sons, pp. 263–86.

Anderson, P. (1999) "Complexity Theory and Organization Science," *Organization Science*, 10(3): 216–32.

Arthur, B. (1994) *Increasing Returns and Path Dependency in the Economy*, Ann Arbor: University of Michigan Press.

Black, L., Cresswell, A., Pardo, T., Thompson, F., Canestraro, D., Cook, M., Luna, M., Martinez, I., Andersen, D. and Richardson, G. (2002) "Dynamic Theory of Collaboration: A Structural Approach to Facilitating Intergovernmental Use of Information Technology," *Electronic Proceedings of the 36th Annual Hawaii International Conference on System Sciences*.

Boland, R.J. (2002) "Design in the Punctuation of Management Action," in Boland, R. (ed.) *Managing as Designing: Creating a Vocabulary for Management Education and Research*, Frontiers of Management Workshop, Weatherhead School of Management, June 14–15, Online. Available at: design.cwru.edu (accessed January 15, 2004).

Cellular Telephone and Internet Association (CTIA) (2003) *Semi-Annual Wireless Industry Survey*, Washington, D.C.

ComCare (2003) The E-Safety Program to Make Americans Safer, Washington, D.C., Online. Available at: www.comcare.org (accessed January 15, 2004).

Connery, N. and Clevenger, C. (2002) *Survey of US Infrastructure: 2000/2050*, New York, NY: ICIS.

Coogan, M. (2000) "A New Product Is Born in Gothenburg," Cambridge Systematics, Inc., *Paradigm Shifts and Fundamental Change in Business and Industry: Case Studies and Descriptive Materials*, File Report-Project J-8B, Transportation Research Board, Transit Cooperative Research Program, Washington, D.C.

Dooley, K (1996) "Complex Adaptive Systems: A Nominal Definition," Online. Available at: www.eas.asu.edu/~kdooley/casopdef.html (accessed January 15, 2004).
Evans, H. (2000) *Water Distribution in Ancient Rome*, Ann Arbor: University of Michigan Press.
Flood, R. (1999) *Rethinking the Fifth Discipline: Learning within the Unknowable*, London: Routledge.
Fountain, J. (2001) *Building the Virtual State: Information Technology and Institutional Change*, Washington, D.C.: Brookings Institution Press.
Frontinus, S. (AD 84) *Strategemata* (intro. book III; *c.* AD 84), Online. Available at: www. www.roman-britain.org/people/frontinus.htm (accessed January 15, 2004).
Fulmer, R. (1997) "The Evolving Paradigm of Leadership Development," *Organizational Dynamics*, 59–75.
Gifford, J. and Pelletiere, D. (2002) "'New' Regional Transportation Organizations: Old Problem, New Wrinkle?" *Transportation Research Record*, No. 1812, 106–11.
Grigg, N., Criswell, M., Fontaine, D. and Siller, T. (2001) *Civil Engineering Practice for the Twenty-First Century*, Reston, VA: American Society of Civil Engineers.
Hall, P. (1998) *Cities in Civilization*, New York: Pantheon, 1998.
Horan, T. (2003) "Integrating the End User into Infrastructure Systems: A Customer Centric Approach to the Design and Function of Intelligent Transportation Systems," *Information Management: Support Systems & Multimedia Technology*, Hershey, PA: Idea Publishing.
Horan, T. and Schooley, B. (2004) "Interorganizational Emergency Medical Services: Case Study of Rural Wireless Deployment and Management," *Information Systems Frontiers* (forthcoming).
Institute for Civil Infrastructure Systems (ICIS) (2000) "ICIS Workshop of Civil Infrastructure Systems Education," Executive Summary, Online. Available at: www.icisnyu.org (accessed January 15, 2004).
Ivory, Y.M. and Hearst, A.M. (2001) "The State of the Art in Automating Usability Evaluation of User Interfaces," *ACM Computing Surveys* (33: 4) December, pp. 470–516.
Johnson, N. (1999) "Recruiting and Retaining Information Systems Staff for Information Age Reform," in Heeks, R. (eds) *Reinventing Government in the Information Age. International Practice in IT-Enabled Public Sector Reform*, London: Routledge, p. 351.
Katz, D. and Kahn, K. (1978) *The Social Psychology of Organizations*, New York: Wiley.
Klein, J. (1994) *Interdisciplinary Studies Today*, Indianapolis, IN: Jossey-Bass Publishers.
Lattuca, L. (2001) *Creating Interdisciplinarity*, Nashville, TN: Vanderbilt University Press.
Marie, J. (2001) "Emerging Trends in European Public Transport – Intermodalism: Facilitating Modal Transfers," Paper presented to ENO Foundation, December.
Markus, M.L., Majchrzak, A. and Gasser, L. (2002) "A design theory for systems that support emergent knowledge processes," *MIS Quarterly* (26: 3), pp. 179–212.
National Emergency Number Association (NENA) (2003) *Report Card to the Nation*, Washington, D.C.
National Research Council (1995) *Engineering Education Designing an Adaptive System*, Washington, D.C.: National Academy Press.
Park, J. and Chandramohan, P. (2004) "Static vs. Recovery Models for Survivable Distributed Systems," *Electronic Proceedings of the 37th Hawaii International Conference on System Sciences.*
Perrow, C. (1999) *Normal Accidents: Living with High-Risk Technologies*, 2nd edn, Princeton, NJ: Princeton University Press.
Pfeffer, J. and Salancik, G. (1978) *The External Control of Organizations*, New York: Harper and Row.
Scott, R. (1981) *Organizations. Rational, Natural, and Open Systems*, Englewood Cliffs, NJ: Prentice-Hall.

Senge, P. (1990) *The Fifth Discipline: The Art and Practice of the Learning Organization*, London: Century.
Simon, H.A. (1996) *The Sciences of the Artificial*, 3rd edn, Cambridge, MA: MIT Press.
Sparrow, R. (2001) "Evolving Knowledge Requirements of America's Civil Infrastructure Managers," *Public Works Management and Policy*, Vol. 5, No. 4, pp. 297–307.
Stacey, R. (1993) *Strategic Management and Organisational Dynamics*, London: Pitman.
Sussman, J. (1995) "Educating the 'New Transportation Professional,'" *ITS Quarterly*, Washington, D.C., Summer.
Sussman, J. and Dodder, R. (2002) "The Concept of a 'CLIOS' Analysis Illustrated By The Mexico City Case," Prepared for ESD Internal Symposium, *Urban Structure*, 79–153.
TRANSCOM (2003) *Program Summary*, Jersey City, New Jersey.
Volpe National Transportation Systems Center (1999) *Building Professional Capacity in ITS: Documentation and Analysis of Training and Education Needs in Support of ITS Deployment*, ITS Joint Program Office, Washington, D.C.: USDOT.
Walls, J.G., Widmeyer, G.R. and El Sawy, O.A. (1992) "Building an Information System Design Theory for Vigilant EIS," *Information Systems Research* (3: 1), March, pp. 36–59.
Wang, M. and Kettinger, W. (1995) "Projecting the Growth of Cellular Communications," *Communications of the ACM*, Vol. 38, No. 10, October, pp. 119–22.
Wenger, E., McDermott, R. and Snyder, W. (2002) *Cultivating Communities of Practice: A Guide to Managing Knowledge*, Harvard Business School Press, Cambridge, MA: Harvard Business School Press.
Winner, L. (1977) *Autonomous Technology: Technics Out of Control as A Theme in Political Thought*, Cambridge, MA: Riverside Press.
Yao, J. and Roesset, J. (2001) "Suggested Topics for a Civil Engineering Curriculum in Infrastructure Management," *Public Works Management and Policy*, Vol. 5, No. 4, pp. 308–17.

CHAPTER FOUR

Policy Perspectives

Jonathan Gifford

Introduction

Since the terrorist attacks in the US of September 2001 the world has withstood a series of wrenching, violent events that have struck at the heart of the infrastructure services that support our way of life.

- The US and its allies launched Operation Enduring Freedom in Afghanistan in October 2001, and now confront a significant infrastructure rebuilding effort
- Sniper shootings along the roadways in greater Washington, D.C., killed ten and injured three (October 2002)
- The US Space Shuttle Columbia disintegrated on re-entry to the atmosphere and killed its crew (February 1, 2003)
- SARS (Severe Acute Respiratory Syndrome) broke out in Asia and spread to Canada (March to July 2003), sickening more than 8,000, killing 774, and stifling global air travel (World Health Organization 2003)
- The US and its allies launched Operation Iraqi Freedom in March 2003, and handed over power to the interim Iraqi government in June 2004, along with the responsibility of rebuilding Iraq's infrastructure
- A power blackout in eastern North America affected roughly 50 million people (August 14, 2003) (US-Canada Power System Outage Task Force 2003)

- An Internet invasion of the W32.Sobig.F virus and the W32.blaster and W32.welchia worms slowed or disabled computer systems worldwide.

Weather has stressed our infrastructures as well; eastern North America experienced a harsh winter in 2002/2003, followed by a very late summer, with torrential rains, and then Hurricane Isabel (September 18–19, 2003), which prompted a first-time system-wide shutdown of the Washington, D.C. Metro and power outages of more than a week in many parts of the US capital. Europe and western North America experienced unusually hot weather in summer 2003, with record temperatures in Europe, and 11,000–15,000 dead from the heat in France alone ("France Heat Wave Death Toll Could Reach 12,000" 2003; Saboun 2003). Wildfires in six southern-California counties consumed 275,000 acres, killed more than 20 and destroyed 3,650 homes and structures, and caused the closure of major highways and airports (Hoffman and Perry 2003).

The August 2003 blackout precipitated a flood of news stories, feature articles, prognostications, congressional hearings, editorials, and of course, policy proposals from those wishing to use the event to leverage their particular interest or point of view. An article in the *New York Times* drew a parallel with the demands on the railroad network during World War I and the operational breakdown and government takeover that followed (Schiesel 2003).

Clearly, infrastructure has played a central role in many of these events. Failed infrastructure precipitated events in some cases, and infrastructure systems usually played an important role in response and recovery. Society depends on its infrastructure systems. When they fail, social well-being is compromised. And society's response to crises – whether induced by infrastructure failures or not – relies on its infrastructure systems. In today's environment, these infrastructures have information technologies (IT) at their core.

While IT may be at the core of contemporary infrastructure, it does not necessarily follow that there is a coherent public policy to support the technological integration of infrastructure systems.

The particular focus of this book is the embedding of information technologies into infrastructure systems. What are the impacts of such digitization? What are the barriers to its occurrence? The incentives? How has it played out in different infrastructure sectors?

This chapter addresses the public policy issues raised by the prospect of infrastructure digitization and what role public policy does and should play in this regard. Public policy, as the term is used here, refers to the constitutional documents, statutes, regulations and court decisions that govern the actions of governments, corporations and private individuals (Dunn 1994). It therefore does include the formal institutions of a society – its international, national and subnational units of government, for example. It does not include the culture, habits and beliefs of a society, which can exert a strong influence on infrastructure systems. There is a gray area, such as the persuasive power of an elected leader, which may not carry any coercive power, but which can still influence behavior.

Concern about public policy arises at least in part from our heightened concerns about and reliance on our infrastructure systems, and the role of IT in promoting confidence in these systems. To what extent has public policy contributed to the digitization of infrastructure systems? To what extent does or should public policy condition and constrain the digitization of infrastructure systems? What other forces shape that digitization? Should public policy press more, or less, or in a different direction? Should it aim to promote, to shape, to impede, or simply to be neutral?

The chapter first discusses some of the drivers and constraints on infrastructure digitization. A second section follows with an illustration in the road transport sector. The third section outlines a conceptual framework for the roles public policy can play in fostering or restraining digitization. The fourth section discusses a range of public policy concerns that arise out of digitization. A final section concludes with observations about how policies can be used to influence the use of IT for infrastructure improvements and what precautions are needed to accomplish this effectively.

Drivers of and Constraints to Digitization

The very term "digital infrastructures" reflects the growing importance of IT in the operation of infrastructure systems. The *drivers* of digitization are several.

Improved technology and systems. The technical capacity of sensors, communication and control systems has improved significantly in recent years (see Chapter 2). Such technical capacity affords significant opportunities for infrastructure owners to implement improved communication and control systems.

Growth of enterprise-wide systems. One form such implementations take is the deployment of enterprise-wide information systems (often called enterprise resource planning, or ERP systems), marketed by vendors such as SAP, Oracle and PeopleSoft. Such systems have become increasingly common in medium to large private and public agencies. While costly and often painful to implement, many agencies have found such systems to improve information sharing, coordination and control (Austin *et al.* 1999; Davenport 1998; Davenport 2000; Deutch 1998; Mabert *et al.* 2001). The implementation of such systems typically involves a tight integration of information sharing across an entire enterprise, including its suppliers (or key suppliers) and its customers or clients.

Replacement of obsolete systems. As aging communication and control systems come due for replacement, infrastructure owners may opt to upgrade to digital systems, since they are simpler and more cost-effective to introduce during replacement or reconstruction than as an add-on or overlay. A potential advantage of digitizing is that it may enable services to meet increasing demands.

One early prognosis of the August 2003 Blackout, for example, was the need for modernizing the power grid. Energy secretary Spencer Abraham has cited cost estimates of US$50 billion over the next decade (Iwata and Kiely

2003). This modernization would entail replacing the electromechanical switches that are currently most widely used, with digital switches that would be more effective for monitoring and controlling current flows in the network.

It is not always the case that such upgrades reduce overall costs, however. Initial outlays can be substantial, and benefits may not be realized until far into the future, and may not live up to expectations. On the other hand, other dimensions of performance besides costs may improve, such as quality and reliability.

Demand pressure. Demand for infrastructure services is highly distributed spatially, and IT offers the opportunity to connect them. Secular increases in demand derive from overall economic growth. Other demand increases arise from the redistribution of demand due to deregulation and other economic policy changes. Deregulation of a wide range of infrastructure supply and user industries has brought sharply different patterns of supply and demand. Much has been made of the electricity disruptions in California in 2000, but similar disruptions remain apparent in industries such as trucking, airlines and drinking water supply.

Demand pressure may induce infrastructure owners or operators to seek options to expand capacity or otherwise improve operational performance through installation of digital systems.

Convergence of infrastructure systems and services. With deregulation has come an erosion of traditional barriers to integration across infrastructure facility and service "stovepipes." Freight shipping, for example, has seen the emergence of third party logistics (3PL) firms that sell integrated "end-to-end" services. Under such services, a shipper of a shipment contracts with a single party, who in turn contracts with different suppliers to provide different stages of a shipment (pickup from origin to first staging area, trunk movement, delivery from final staging area to delivery point). This service includes coordination of all transfers, including management of responsibility for loss or damage. Such integration would have been quite difficult before deregulation, and heightens the value of system integration and associated digitization (Slack 2001).

Heightened attention to critical infrastructure protection (CIP). In 1998, Presidential Decision Directive 63 established national policy on critical infrastructure protection, with a primary emphasis on the security of the nation's information infrastructure (US President 1998). Subsequently, concern about critical infrastructure protection has expanded to include the security of the facilities and institutions that comprise the nation's critical infrastructure. As attention to CIP has increased, infrastructure owners and operators have utilized IT to enhance protection and security.

Public expectations. As technical capability for communications and control has improved, public expectations about the performance of infrastructure owners and operators have increased as well. The public and the media expect real-time information on system outages and service restoration schedules.

Limitations on physical capacity expansion. Many infrastructure owners and operators face high costs and significant delays when expanding existing facilities or adding new links or nodes to keep pace with growth in demand. Important sources of delay are the opposition of interest groups, procedural requirements for obtaining public permits and approvals, and acquisition of necessary land or rights of way. Improved monitoring and control offer the potential to increase operational capacity within the confines of existing physical capacity.

Cost pressure. Traditional approaches to providing infrastructure and delivering services over them have become increasingly costly. In some cases, it has become difficult to pass on increased costs, especially to commercial and industrial users. Digitization offers the possibility of significantly reduced costs, although often only after substantial initial outlays.

In addition to these drivers of digitization, infrastructure owners and operators nonetheless face significant *constraints* on their ability to exploit digitization. Many of these are characteristic of the typical challenges to developing information systems in any standard management information systems text (see for example, Alter 2002; Organization for Economic Cooperation and Development 2001). Others are distinctive to infrastructure systems.

Unproven technology. While promising, many of the digital technologies have yet to be proven reliable, a serious limitation in safety critical infrastructure systems. Many are relatively new, and applications to evaluate their viability are just being developed.

Technological obsolescence. The rapid pace of innovation and technological development can render costly IT installations obsolete in a short period of time. In the infrastructure domain, obsolescence may confound expectations about the service life of facilities and equipment.

Inadequate codes and standards. The development of the codes and standards that govern the infrastructure community often requires years. If the cycle of technological innovation outpaces the cycle of code and standard development, infrastructure owners may be placed in a situation where they are specifying obsolete technology for their system improvements because codes and standards for the more current equipment have often not kept up, and in many cases are not yet available.

Allegiance to traditional business practices. Infrastructure owners and operators may encounter resistance to the implementation of IT-based systems due to allegiance by their workforce to traditional business practices. Resistance to change is not limited to blue collar workers, and may be exacerbated by labor agreements. Moreover, pink- and white-collar workers up to and including senior management and board members may not be open to utilization of IT-based systems.

Procurement requirements. Public infrastructure owners are often subject to procurement procedures and requirements that are poorly suited to IT procurements.

Requirements to accept low bids, for example, can lead to significant problems, both in initial procurements and in long-term operations and maintenance (Kelman 1990; Kelman 1999). Furthermore, any system of procurement that requires clearing of funds between public entities adds additional complexities. Public agencies have financial controls that frequently lead to difficulties because one agency's procedures do not integrate smoothly with another's. Something as simple as different accounting periods (weeks versus months, for example) can lead to extremely high transaction costs. Agency financial controllers are often highly resistant to any changes in their control protocol out of (often legitimate) concerns about discharging their responsibilities to maintain careful control of public taxpayer funds.

Cost and schedule uncertainty. Costs and schedules for IT-based systems are notoriously difficulty to estimate. Costs for both the inputs to system development and the development process itself are subject to high levels of uncertainty. Cost overruns and schedule slippage can generate "bad press" and tarnish the reputation and standing of both public and private infrastructure owners.

Requirements "creep." IT-based systems often follow a development model that begins with the specification of user requirements, which then serve as inputs into a system design. However, in the development of wholly new IT-based systems, users are often not in a position to specify their requirements in a way required by system developers. Users may be unfamiliar with the capabilities of IT-based systems, and hence unable to formulate clear, unambiguous requirements. Requirements also change in the course of a system's development, for a variety of reasons. Sometimes users modify their requirements as they become familiar with what systems can do. Sometimes changes derive from changes in the environment of the organization. For example, security emerged as a major new requirement for many infrastructure owners and operators after September 2001. Such changes in system requirements can be extremely damaging to a system development process, both in terms of schedule and cost.

This brief cataloging of drivers and constraints emphasizes the fact that infrastructure digitization can pose significant challenges to owners and operators, and sets the stage for the examination of policies.

Illustration: Policies Affecting Digitization of Road Transport

Given these drivers and constraints on infrastructure digitization, what is and should be the role of public policy? Should it seek to promote? To deter? To use the strategic management argot, should it row (provide motive power)? Should it steer (guide and direct)?

As an illustration, consider one infrastructure system where digitization has increased substantially in recent decades – road transport. Twenty years ago, concepts of information systems in road transport were in their infancy. Computer models were well established for planning and analysis, but the road transport sector was virtually devoid of the use of IT for the management and operation of transportation systems. By contrast, air transportation and some new rapid transit systems, such as San Francisco's BART and the Washington (D.C.) Metro system, used such systems extensively.

Twenty years hence, the nexus between management information systems (MIS) and transportation systems is extensive. Improvements in sensors and communication technologies now enable integrated monitoring and coordination in real time. The field of intelligent transportation systems (ITS) has exploded and become an essential tool for many transportation operators (See Chapter 6 for a further discussion of the development of ITS).

Experience during Hurricane Isabel in 2003 exemplifies the extent to which digitization has already occurred. Only a few years ago, truck communications systems relied heavily on CB radio and land telephone lines. Today, they depend on GPS, cellular telephones and digital PBXs. Trucking and logistics firms were seriously disabled by the power failures after Isabel. Many dispatchers lost track of their trucks. Mobile equipment was able to transmit, but it could not go anywhere. GPS transponders on trucks were working, but the dispatch centers were without power to receive them. Cellular phones failed to work because cellular towers were without power. PBXs were out of service, and analog telephone lines were in high demand since they continued to work in the absence of power (Legler 2003).

What has been the role of public policy in this transformation? On the one hand, federal policy has sought aggressively to promote ITS by supporting research and development, field tests, development of a national ITS architecture and standards, and evaluation. Funding has substantially exceeded US$1 billion since 1991.

By some accounts, federally sanctioned planning requirements are poorly suited to ITS investments because their underlying public policy requires a 20-year planning horizon and understates the importance of qualitative improvements that are often the primary result of ITS investments, such as reliability and improved information quality.

Government procurement requirements, another dimension of public policy, have also presented challenges for government bodies seeking to purchase ITS facilities and services. A tradition of low-bid oriented purchasing policies has proved inappropriate for the complex, multidimensional products and services embodied in ITS.

States, and to a lesser extent local governments, own and operate the vast majority of highways where ITS has the highest potential payoff. Obviously, the public policies of these entities, as reflected in their explicit budget priorities for ITS over other forms of investment, as well as their procurement and contract management capabilities, have an important impact on the adoption of ITS products and services.

Another major stakeholder in the deployment of ITS is the vehicle manufacturers and their regulatory body, the National Highway Traffic Safety Administration

(NHTSA). NHTSA has the authority to mandate particular technologies on the basis of safety. The US Department of Transportation has also undertaken vehicle-oriented initiatives (e.g. the Intelligent Vehicle Initiative, or IVI).

Yet vehicle manufacturers rather than public policy have played a dominant role in advancing the development of vehicle-based digitization initiatives. The American tort system may be the most potent dimension of the public policy environment for vehicle-based systems, since concern about liability for the failure of innovative systems has in some cases held back digitization initiatives.

Finally, public policy plays an important regulatory role in the motor carrier industry, at the federal, and especially at the state level. States are responsible for issuing permits for motor carriers who operate in their jurisdictions and are responsible for inspecting vehicles to ensure compliance with safety requirements and conformity requirements on weight and "placarding" (i.e. displaying signs on trailers indicating whether contents are hazardous).

This example illustrates a range of public policy roles and responsibilities:

- research and development support
- deployment and field test support
- evaluation
- system architecture and standards support
- financial incentives and support
- planning and procurement requirements
- infrastructure owner and/or operator
- vehicle manufacturing regulator
- motor carrier operator regulator.

This list is a useful point of departure for considering the role of public policy more generally.

Conceptual Framework for Public Policies and Digitization

In considering public policy roles and responsibilities in infrastructure digitization, it is useful to distinguish between a system's "core technology" and its "control structure." Core technology refers to the intrinsic technology of a particular infrastructure system, such as a rubber-tired vehicle running on pavement, or the flow of electricity across a network, or the passage of an aircraft from taxi to takeoff, to en route, to approach, to landing to taxiing, or the passage of data packets over the Internet. The role of information technology in the core technology varies from infrastructure system to infrastructure system. In the case of tire-on-pavement, the "intrinsic" level of digitization is limited – although "smart tires" or anti-lock braking systems could have some relevance. In the case of ancient technologies such as ocean sailing vessels, the intrinsic level of digitization is essentially nil. In the packet-across-Internet, on the other hand, the intrinsic level of digitization is high.

The communication and control structure that governs a network may also vary in its degree of digitization. How much do the operators of the network know about its condition? What are their sources of information? What is the lag and cycle time of their information? What are the respective roles and responsibilities of the system's operators and users, that is, how responsible (and accountable) are they for system performance?

One general observation is that the extent of digitization of the control structure could be quite sensitive to public policy. Public policy can also affect knowledge about a core technology and its rate of deployment. The digitization of the core technology, however, is more intrinsic to an infrastructure system itself. Like it or not, government cannot change the laws of physics.

Consider that the core technology of an infrastructure system typically exploits some electro-mechanical properties. This core technology is not itself a product of public policy but rather of "hard" science and engineering. By directing research, development, deployment and field tests, public policy can affect the rate at which such core technologies come to be discovered, understood and ultimately deployed. Military support for aircraft technology development is an example.

Public policy has much more influence over the control structure, however. Public policy can dictate or influence who is accountable for system performance, and the consequences of performance shortfalls. Such accountability and consequences can have a strong influence on the level of investment in the control structure, and the extent of control necessary for a core technology to be released for public use.

Potential roles for public policy, adapted from the list generated in the foregoing discussion of road transport, include:

- research and development support
- deployment and field test support
- evaluation
- system architecture and standards development
- financial incentives and support
- planning and procurement requirements

- tort and liability policy
- infrastructure owner and/or operator
- infrastructure regulator (e.g. public utilities commission)
- user-owned equipment owner and/or operator
- user-owned equipment regulator.

Policy Concerns and Uncertainties

In addition to the roles that public policy can play in shaping the digitization of infrastructure systems are a set of public policy concerns about the impact of such digitization. Society depends critically on its infrastructure systems to provide services and materials. These concerns include reliability, economic efficiency, interdependence, accountability, equity, data retention and privacy, safety and security, technological obsolescence, organizational obsolescence, regulatory lag, and education and training.

Reliability. The growth of digitization raises important questions about the reliability of infrastructure systems, which is usually very important to users. But expectations about reliability vary considerably across networks, as do the consequences of failures. As mentioned earlier, the move from CB radio and land line telephony to cellular telephony in the trucking industry created a vulnerability that was exposed by Hurricane Isabel.

As the economy has shifted towards just-in-time inventory management, the consequences of supply chain disruptions have increased. Supply chains are longer. Companies are increasingly relying on "pull" logistics rather than traditional "push" logistics systems, which are highly dependent on integrated IT systems and reliable transportation.[1] Service delivery has also become more dependent on highly perishable supplies, which exacerbates the impacts of supply chain disruptions. For example, many radiological diagnostic and treatment supplies have shelf lives of 12 hours or less and are produced and shipped overnight to hospitals for use within 24 hours. Fresh food deliveries with short shelf life like rice balls and packed lunches are now delivered three times per day in Japan ("*Konbini* Culture Hits Asia" 2003). Disruption of such supply chains can impair service delivery in ways that cannot be hedged against through increased inventories.

Public policy to ensure reliability is tricky, however. Telephone and power networks as public utilities have traditionally been subject to regulation of their reliability. With deregulation of both, and the associated increases in competition, reliability becomes somewhat more problematic. The reliability of the highway system has not traditionally been a subject of explicit public policy, but has become an increasing concern with tightly integrated supply chains. Reliability of air traffic control for maintaining separation of aircraft, on the other hand, has been an explicit design objective. Air service reliability in terms of schedule adherence, on the other hand, has been largely left to the airlines. Public policy has relied primarily on provision of on-time performance statistics.

Economic efficiency. Some situations afford the possibility of providing users choice about the level of services they acquire. "Congestion pricing" or "value pricing" in highway transport gives users an option, for a price, to avoid congestion. Electric power suppliers have begun to offer "interruptible" service to users at a discount. Some forms of failures are unacceptable, however, such as mid-air collisions. Such systems require sophisticated metering and control systems

Disruptions to infrastructure services because of increased reliance on IT could potentially have significant economic impacts. In the aftermath of the September 2001 terrorist attacks, supply chains across the Canadian border to auto manufacturing facilities in Michigan were disrupted by border closures. Reportedly, urgent calls to the White House narrowly averted shutdowns. Total logistics costs in the economy have dropped substantially over the last two decades, from 14 percent to ten percent of overall GNP, with the largest declines in expenditures on inventory.[2] The economic consequences of severe disruptions to freight supply chains, or a move back towards traditional higher-cost inventory and warehouse systems could damage economic productivity and prosperity.

Interdependence. Closely related to the issue of reliability is the interdependence between infrastructure systems, such as transportation and telecommunications or power and telecommunications. Of course, when one infrastructure service relies on service inputs from other infrastructures, its performance levels are a joint product of multiple sectors. IT can increase interdependency, yet at the same time can support brokering the interdependencies among other infrastructures.

Accountability. Who is responsible for system failures that arise out of such joint services and the losses that they may give rise to? This is not only a question of legal liability, but also public accountability.

Equity. Equity considerations are not necessarily a problem at the level of the individual end user, as the digital content of an infrastructure is likely to be invisible to the end user in many cases. However, cross subsidies are an issue. If a digitally enabled service is useful to only a subset of the users of an infrastructure service, how should it be priced and paid for? In the public utility domain, this has been a matter of great concern. In the telephone service area, for example, enhanced services such as call waiting and voice mail have typically been required by regulators to be offered at premium prices. But allocation of the costs of developing and providing such services can be arbitrary, and fair pricing of such services can become a matter of contention in a regulated environment.

Data retention and privacy. The rise of digitization raises a number of important issues about data retention and privacy. As messages and control move increasingly to digital delivery, the capacity to store and retain records increases. In a public agency setting, such records may become subject to the Freedom of Information Act or other public access laws. In the event of a criminal investigation or civil suit, such records can be subject to subpoena. Such records may contain private or sensitive information about private individuals using the infrastructure system, or about employees or contractors involved in operations. Public policy requirements for retention and disclosure could become a concern, as has already become the case with medical records (Gifford and Carlisle 2004).

Safety and security. Public safety and security are longstanding concerns of public policy, and they bear particularly on digitization of infrastructure systems. Some safety and security issues arise from reliability and interdependence, as discussed above. Infrastructure system vulnerability can increase as a result of digitization, due to both intentional and unintentional actions. Security from hacker attacks are important, especially if injury or loss of life are at issue. As critical infrastructure protection has risen in the public consciousness, the complexity of trying to maintain security in an open society where many of the infrastructure assets are privately owned and operated has become increasingly evident. Digitization is an important dimension of this. Real time monitoring and control systems afford the possibility for public seizure of information and, possibly, control. Control and communication systems are largely privately provided in the US, yet their security has enormous public policy importance.

Technological obsolescence. As indicated earlier as a constraint, the pace of technological innovation in IT and communication has been very rapid in the last several decades. Along with new capabilities has come a challenging problem of obsolescence. Legacy systems and technologies are expensive to maintain and expensive to replace.

Organizational obsolescence. Legacy systems embody not only hardware and software, but also "org-ware" – that is, business processes and organizational arrangements that are often also very difficult to re-engineer as market conditions evolve. Use of the term "legacy airlines" to refer to traditional major carriers in the US industry is now common. In that case, evolution of airline business practices and business models has been so constrained by labor agreements and insurance and financing provisions for aircraft, some argue, that their viability is now seriously in jeopardy.

Regulatory lag. Closely related to the pace of technological breakthroughs is the set of laws and regulations that govern such systems. Laws and regulations can easily be outpaced by technological advancements. For example, a US district court recently enjoined the Minnesota Public Utilities Commission from regulating the voice-over-Internet company Vonage, citing the US Congress's decision to allow the Internet to remain free from undue regulation (Vonage Holdings Corp. v. Minnesota PUC, 2003). The public policy concern, then, is to monitor recent and emerging breakthroughs to identify where laws and regulations require modification. Regulatory review and revision can be particularly resource intensive for rulemaking agencies, and so prioritizing where regulatory review is most beneficial is also important.

Education and training. As digitization has progressed in various infrastructure sectors, it has brought particular challenges to the education and training establishment. Educational curricula need to remain current, and training the existing workforce to provide requisite knowledge and skills in the information and communication technology sector is critical. The special mix of skills is particularly challenging: to combine a knowledge of infrastructure, a knowledge of digital technology, how to think about their combinations, and the public policy issues they raise. This has led to a call in the transportation field, for example, for a "T-shaped" transportation professional, with breadth across a range of infrastructure-related fields representing the horizontal cross bar of the "T," and depth in a particular specialty representing its vertical stem (Sussman 1995). Part of this educational approach includes understanding how public policy can be formulated to support the integration of IT and infrastructure.

This listing of public policy concerns belies the reality that for public policy to be effective, it would need to address the technological dimension in a coherent and consistent manner. Yet, the fragmented nature of infrastructure policy makes the connection between public policy and IT for infrastructure highly disbursed, even "hit or miss" in some circumstances.

Concluding Remarks

Public policy can play an important role in fostering the adoption of information and control technologies in infrastructure systems. This chapter has identified eleven possible roles, ranging from the sponsorship of research and development to the direct adoption of such systems as the owners and operators of an infrastructure system. The appropriate combination of such policies will vary among infrastructure systems and also as matters of national policy traditions. The challenge for policy makers and analysts, then, is to craft public policies that can affect the deployment of IT in infrastructure systems in a timely, efficient, and socially defensible manner.

The adoption of IT in infrastructure systems also raises a host of public policy concerns, from privacy to education and training. The pace and direction of such adoption will vary among systems and across nations, and thus the urgency of these concerns and the policy remedies that may be introduced will vary as well.

Yet both the promise and the peril of digital infrastructures remain potent. The challenge for infrastructure owners and operators, as well as public policy makers, is to exploit the promise and manage the peril. It is a challenge well worth meeting. But as the list of calamities that opened this chapter illustrates, these challenges may sorely test the capabilities of new digital infrastructure professionals.

Acknowledgments

W. Homer Carlisle provided invaluable research assistance for the completion of this chapter. Remaining errors are the responsibility of the author.

Notes

1 "In a push logistics system, suppliers ... deliver materials to an integrated manufacturer, whose manufacturing process pushes products ... to a distributor or retailer and ... thence to the customer. To dampen swings in demand and supply..., suppliers, manufacturers, distributors, and customers maintain large inventories of critical materials and products. By contrast ... a pull logistics system depends heavily on timely and accurate information about customer demand to track market movements and fast and reliable transportation to meet customer demand. Point-of-sale data ... tell the distributor, manufacturer, and supplier how much product is needed when and where. But the process is growing more complicated as ... products flow back up the supply chain for recycling; second and third tier global suppliers ... are added; integrated manufacturers give way ... to virtual companies; and 3PLs ... are added to manage the freight flows. The complexity and interconnected-ness of a "neuro-logistics" system is both its strength and its weakness" (Caldwell n.d.).

2 Cass/ProLogis, *10th Annual State of Logistics Report* (1998), cited (Caldwell n.d.).

References

Alter, S. (2002) *Information Systems: Foundation of e-Business*, 4th edn, Upper Saddle River, NJ: Prentice Hall.

Austin, R.D., Escalle, C.X. and Cotteleer, M. (1999) *Enterprise Resource Planning, Technology Note*, Cambridge, MA: Harvard Business School Publishing.

Caldwell, H. (n.d.) I-Freight: "The Role of the Interstate Highway System in Freight Transportation," PowerPoint presentation, *Interstate Vision Initiative* US Federal Highway Administration, Online. Available at: ops.fhwa.dot.gov/freight/pp/Caldwell-I-Vision_Presentation_v6.ppt (accessed October 17, 2003).

Davenport, T.H. (1998) "Putting the enterprise into the enterprise system, reprint 98401," *Harvard Business Review*, July–August, pp. 121–31.

Davenport, T.H. (2000) *Mission Critical: Realizing the Promise of Enterprise Systems*, Cambridge, MA: Harvard Business School Press.

Deutch, C.H. (1998) "Software that can make a grown company cry," *New York Times*, November 8, p. 31.

Dunn, W. (1994) *Public Policy Analysis*, 2nd edn, Englewood Cliffs, NJ: Prentice-Hall.

"France heat wave death toll could reach 12,000," (2003) *Canadian Broadcasting Corporation* (August 31), Online. Available at: www.cbc.ca (accessed December 5, 2003).

Gifford, J.L. and Carlisle, W.H. (2004) "Data retention & access regimes for wireless message logs in the US: an exploratory analysis," *Transportation Research Record*.

Hoffman, A. and Perry, T. (2003) "Southern California Fires: Preventing Another Catastrophe May Take Billions, Panel Is Told," *Los Angeles Times*, December 5, B1.
Iwata, E. and Kiely, K. (2003) "Grid needs $50 bil. makeover, chief says," *azcentral.com*, November 20, Online. Available at: www.azcentral.com/news/articles/1120blackout-qampa20.html (accessed December 5, 2003).
Kelman, S. (1990) *Procurement and Public Management*, Washington, D.C.: AEI Press.
Kelman, S. (1999) "White House-initiated management reform: implementing federal procurement reform," in Pfiffner, J.P. (ed.) *The Managerial Presidency*, 2nd edn, College Station, TX: Texas A&M University Press, pp. 239–64.
"*Konbini* culture hits Asia," (2003) *Look Japan*, February, Online. Available at: www.look-japan.com/LBecobiz/03FebEF.htm (accessed December 5, 2003).
Legler, J. (2003) personal communication (October, 2003).
Mabert, V.A., Soni, A. and Venkataramanan, M. (2001) "Enterprise Resource Planning: Common Myths vs. Evolving Reality," *Business Horizons*, May 15.
Organization for Economic Cooperation and Development (2001) *The Hidden Threat to e-Government: Avoiding Large Government IT Failures*, PUMA Policy Brief, 8, Online. Available at: www.oecd.org/puma/Risk/ITfailuresE.pdf (accessed August 31, 2003).
Saboun, F. (2003) "France heat wave death toll set at 14,802," *Washington Times*, September 25.
Schiesel, S. (2003) "In frayed networks, common threads," *New York Times*, August 21, G1.
Slack, B. (2001) "Intermodal transportation," in Brewer, A.M., Button, K.J. and Hensher, D.A. (eds) *Handbook of Logistics and Supply-Chain Management* Chapter 9, *Handbooks in transport*, Vol. 2, Amsterdam: Pergamon. pp. 141–54.
Sussman, J. (1995) "Educating the new transportation professional," *ITS Quarterly*, 3(1).
US President (1998) Critical infrastructure protection, Presidential Decision Directive, No. 63, May 22.
US-Canada Power System Outage Task Force (2003) "Interim Report: Causes of the August 14th Blackout in the United States and Canada," Online. Available at: https://reports.energy.gov/814BlackoutReport.pdf (accessed December 10, 2003).
Vonage Holdings Corp. v. Minnesota PUC (2003) Civil No. 03-52287 (MJD/JGL), Trans., *D. Minn*, October 16.
World Health Organization (2003) "Summary of probable SARS cases with onset of illness from 1 November 2002 to 31 July 2003," September 26, Online. Available at: www.who.int/csr/sars/country/table2003_09_23/en (accessed December 5, 2003).

Part II
Infrastructure Sectors

CHAPTER FIVE

Water

Rae Zimmerman

Introduction

The history of the digitization of water infrastructure technology parallels the history of the growth of human settlements. At one time, the drinking water and sanitation needs of civilization relied on gravity systems with only the occasional assistance of mechanical systems to overcome topographical irregularities. Over the course of the twentieth century, as population grew worldwide, the provision of water supply and wastewater management services evolved from being largely independent, self-reliant entities to relatively more interconnected and centrally controlled systems often spanning hundreds of miles. Information technologies (IT) have been increasingly introduced to bridge the spatial divide between the location of water supply and demand, and to meet changing values such as increased reliability, water quality, and security.

Potable water and wastewater systems are separate systems, yet they often share resources, technologies, markets, and regulatory philosophy. The quality of potable water, for example, often depends upon the quality of wastewater where both use the same natural waterways. In addition, the two systems have similar interdependencies with other infrastructure systems and with one another (Gillette *et al.* c2002). These similarities support examining the role of IT in these systems together.

This chapter addresses the increasing reliance of water infrastructure on IT. It draws from a very diverse literature encompassing water resources, product development in IT, and the science of detection. First, the growing dependency of water systems on IT is portrayed in terms of water demand, supply, and technology, and public values and expectations. Second, the technology to support IT applications for water infrastructure is evaluated in terms of its ability to improve supply and demand imbalances. Cases are presented that illustrate the interconnections between information and water technologies in light of user needs. Finally, the concluding section provides new directions for combining knowledge of IT and water systems to improve the quality of water services.

Drivers of IT Usage in the Water Industry: Resources, Technology, and Customer Needs

Mobilizing Resources to Meet Needs – Bringing Water Resources to People

Water Supply

Water supply has always been abundant, but not necessarily accessible to people. Bringing water to people has been accompanied by new technologies, many of which rely increasingly on IT to track the resources, the users, and to support the systems that bring water often many hundreds of miles to its point of consumption. Without these new technologies, this would likely be difficult to do manually or even mechanically. Three factors reflect the magnitude of the water supply distribution dilemma – water availabilability, use, and need.

Availability

Though water resources are abundant, they are unevenly distributed, which affects their capacity to meet public water supply needs. The global availability of water is estimated at about 332.5 million cubic miles (1.386 billion cubic kilometers) (UNESCO 2003: 68), but about 97.5 percent is salt water (Wright and Nebel 2002: 214). Only about one-third of the freshwater component is available since the rest is ice (Wright and Nebel 2002: 214). Like total water, freshwater is not evenly distributed either globally or locally. For example, fresh water availability per capita per year ranges from over 2.64 billion gallons (10 million cubic meters) in Greenland to 2,641 gallons (10 cubic meters) per capita per year in Kuwait (UNESCO 2003: 70–4). Sanitation problems can dramatically affect the availability of water supply by impairing the quality of drinking water where the two systems rely upon contiguous waterways.

Use

By the end of the twentieth century, freshwater use worldwide was estimated to be 1.3 trillion gallons annually (4.9 trillion liters or 5,000 cubic kilometers) (UNESCO 2003: 13) or about 3.6 billion gallons daily (13.6 billion liters daily). The amount of use is not necessarily related to the size of population. For example, the US withdrawal of freshwater is about ten percent of estimated worldwide use (Solley *et al.* 1998: 6), whereas the US accounts for only 4.5 percent of the world's population (US Bureau of the Census 2003a; US Bureau of the Census 2003b). Over time, water withdrawal rates worldwide, however, have largely paralleled the rate of population growth at least since the middle of the twentieth century (Gardner-Outlaw and Engleman 1997 and UNESCO 2003: 13).

Need

One basis for estimating water quantity is water need, but water need is not a well-defined concept. The concepts of shortage or deficit, access, and demand are commonly used to portray components of need. The United Nations Educational, Scientific and Cultural Organization (UNESCO) uses the concept of shortage or deficit, estimating that of the world's population of 6.1 billion people in 2001 (UNESCO 2003: 12) two billion people in 40 countries experience water shortages. "1.1 billion do not have sufficient drinking water and 2.4 billion have no provision for sanitation" resources (UNESCO 2003: 10; WHO/UNICEF: 2000). The worldwide deficit is expected to increase by the middle of the twenty-first century ranging from two billion people in 48 countries to seven billion people in 60 countries experiencing water scarcity (UNESCO 2003: 13; Gardner-Outlaw and Engelman 1997). Access to potable water is one aspect of need, since it may influence how much water people use. It varies substantially by country, ranging from a low of 41 percent in Rwanda to 100 percent in some North American, European, and Asian countries having access to reliable water supplies. (Esty and Cornelius 2002: 229 also using figures from WHO/UNICEF 2000). Finally, another component of need is the demand for water. Gleick, for example, indicates that water consumption for human survival is met with "3–5 liters of clean water per day" (0.8–1.3 gallons) and 20 liters (5.3 gallons) creates improvements in human health (Gleick 2000: 10). Demand is a fragile concept as is the concept of need, and varies with availability, access and cost.

Wastewater

As in the case of water supply, sanitation services for the treatment of wastewater are becoming more distant from where wastewater is generated, and IT has been playing an important role in linking the two. Also, as with water supply, water quality control has contributed to the increasing use of IT for monitoring, surveillance, and control, particularly as standards for water purity have become more stringent and wasteload generation has increased. Water quality standards traditionally targeted a few chemical, biological, and physical properties of water, namely oxygen deficiency, nutrients, toxins, pathogens, and solids, and now cover a couple of hundred substances. The quality of wastewater generation is reflected in part by US Environmental Protection Agency (EPA) estimates of the amount of oxygen demanding substances, measured as biochemical oxygen demand (BOD), entering publicly owned wastewater treatment (POTW) plants. According to EPA, wastewater generation has doubled between 1940 and 1996 (US EPA 2000: 2–38). UNESCO (2003: 10) cites an estimate of two million tons a day for the amount of human waste entering watercourses. Water quality, measured as the quality of natural waterways rather than discharges into waterways, is highly variable all over the world. One report found that North American and northern European countries were generally ranked as having the highest water quality, whereas countries in Africa and the Far East were ranked as having the lowest quality (UNESCO 2003: 140; Esty and Cornelius 2002).

Water Infrastructure to Manage Product and Process

IT and the processes for the provision of potable water supplies and clean water converge at the location of infrastructure systems, and knowledge of water infrastructure technologies, their goals, and management trends are critical to understanding how IT supports water systems.

Water Infrastructure Technologies

WATER SUPPLY

The modification of water and land to increase access to water resources has occurred for thousands of years. This often results in the use of very large and geographically concentrated infrastructure to manage water for individual users, such as large reservoirs and dams to capture and store vast amounts of water, aqueducts to transmit water, and some 880,000 miles of distribution lines to reach end users (NRC 2002). Many intermediate facilities also exist such as wells and well fields, storage reservoirs, pumps, and treatment plants, all requiring information systems to link them together. Increasing distance between sources and points of consumption usually increases the length of the physical systems that connect them, resulting in the need for greater automation. Configurations vary dramatically. Independent water supply systems in sparsely developed areas typically consist of single isolated systems like wells to tap groundwater or pipes to convey water from surface waters, such as lakes and rivers, possibly with an electrically driven pump to bring the water to the point of consumption and distribute it to a small number of users. Community water supply systems, however, supply many users, and the number of components depends on system size and location of supplies relative to users. Other critical infrastructure systems are required to support these components regardless of their size and complexity, and can influence water system configurations. Electric power, for example, is needed to operate pumps in rural areas as well as for large treatment works, distribution systems, and buildings in large urban areas. Both small and large systems require transportation infrastructure for the provision of raw materials, labor and the conveyance of waste products.

WASTEWATER TREATMENT

The story of water supply management does not end with the consumption of water. Wastewater generated through water use sets in motion another set of technologies. Wastewater management infrastructure at a superficial level shares much in common with water supply infrastructure in that it involves water treatment and an intricate system of networks for water transport.

The response to water quality problems has traditionally been largely technological, with the particular technology used varying with population density. The US national assessment of wastewater treatment by the US Geological Survey (USGS) defines the wastewater treatment process as "the processing of wastewater for the removal or reduction of contained solids or other undesirable constituents" (Solley *et al.* 1998: ix). The facilities are defined as encompassing "facilities engaged primarily in the collection, treatment and disposal of wastewater conveyed through a sewer" (Solley *et al.* 1998: 58). Wastewater treatment begins with a network of collection systems, however, that are far more complex than those for water supply, since unlike water supply, the facilities begin at the individual sources of wastewater generation and converge at a treatment plant. The configuration of collection systems varies depending on the source of the water, e.g. street surfaces, household sanitation systems, or industrial processes. Like large water supply treatment systems, wastewater treatment plants are designed to remove substances in an influent water stream and produce a treated water output that meets regulatory and user objectives. Using USGS and US EPA estimates for wastewater treatment (Solley *et al.* 1998: 61; US EPA 2002: 10), about 16,000 public wastewater treatment plants serve about three quarters of the population, discharging about 41 billion gallons of water per day. The remaining one-quarter of the population remains unconnected to public facilities.

Treatment processes are usually divided into those that physically remove waste material, for example through screening, skimming, and filtering, those that degrade primarily organic material through bacterial action, and those that remove substances, for example nutrients, from wastewater via chemical action. The three processes are generally called primary, secondary, and tertiary respectively because of their approximate placement in the treatment process, though the division is not necessarily a strict one. For example, sand filtration – a physical process – is normally called tertiary. Wastewater treatment plants often differ from one another in the types and combinations of equipment and processes used in them since water quality standards vary geographically and standards typically do not specify design (e.g. Yang and Kao 1996).

The implications of this for IT are that a large number of diverse processes and pieces of water purification equipment need to interface with measurement and communication equipment both inside and outside of the facilities, increasing the variety and complexity of IT support. IT equipment is used to gauge the rate of water flow (hydraulics), reactions for the removal of organic and inorganic substances including nutrients and pathogens, and associated use and behavior of raw or input materials such as energy and oxygen. IT is used not only to measure the various states, characteristics, and operational effectiveness of the water stream as it passes through all of the processes but also to ensure that social and economic goals are met as well. IT systems have to target upsets, anomalies, extremes and other deviations from normal or desired conditions. The challenge is that IT has to continuously adapt to changes in the way regulatory agencies and communities define quality objectives.

Goals for Water Quality: Quality and Security

Quality

Quality goals for water supply and wastewater are reflected in the very large and growing numbers of chemicals regulated. Internationally, standards for water supply are most commonly drawn from the US Safe Drinking Water Act, those in Europe, and guidance provided by the World Health Organization (Kahn *et al.*

2001: 7). The goals of water supply management, originally aimed at protection of public health from disease outbreaks from waterborne pathogens, gradually expanded to protection against diseases with long-term latent effects that often had chemical rather than biological origins. As a consequence, regulated contaminants expanded from an initial eight substances when the Public Health Service regulated water supplies in the early 1960s to a couple of hundred substances by the 1990s. The coverage of standards for water discharged from treatment plants similarly expanded to over several hundred chemicals. In addition to the increase in the number of standards, regulated concentrations of substances in both supply and wastewater discharges have declined in some cases to levels of parts per billion and trillion as detection became more refined. Social goals for water infrastructure also expanded to encompass concerns related to the protection of communities adjacent to water and wastewater treatment plants and workers within the facilities from impacts such as traffic associated with the movement of raw materials and wastes, potential dangers associated with chemical storage, and odors. These social goals resulted in a greater expansion of the number of criteria and standards, and while increasing the demand for monitoring capability, were probably a key factor in the increased digitization of control systems for water infrastructure.

SECURITY

Security concerns have also grown, especially since the September 11 attacks. IT is being evaluated to a greater extent as a means to address those concerns. Though this issue has been discussed primarily for water supply, the technologies and issues are similar for wastewater treatment. The National Research Council (NRC) (2002: 250) recommended that improvements in the monitoring of biological and chemical agents will need "new sensors for better, cheaper, and faster sensing" which "should be small, distributed, resistant to interference, and robust against false positives." Legislation now requires water supply systems above a certain size to have vulnerability assessment plans, many of which rely on IT. Wong (2002: 28–30) describes specific IT needs in water infrastructure facilities such as heightened intrusion detection systems to secure control rooms, electrochemical gas detectors to speed evacuation, and distributed or decentralized systems using Supervisory Control and Data Acquisition (SCADA) to allow separate units to communicate.

IT will have to be tailored to the nature of anticipated security breaches. Though direct attacks on water supply systems are rare, break-ins have occurred and threats are well known ranging from illegal facility access to contamination of the supply, with the former generally more common than the latter. For example, among the small but symbolic illegal water system break-ins in 2003 alone were those reported in Carpentersville, IL (Lissau 2003), Volusia, FL (Local6.com 2003), Willcox, AZ (Cole *et al.* 2003), Grand Rapids, MI (Woodtv.com 2003), Shelton, WA (Kinne 2003), and Montreal, Canada (CTV.ca News Staff 2003). Contamination threats include one by Kurdish rebels involving potassium cyanide to poison the water system of a military base in Turkey in 1992 and a threat by a German biologist to contaminate water supplies with biological toxins in 1973 (Kahn *et al.* 2001: 7). An actual incident was the sabotage of the Ta' Kandja water galleries in Malta involving the intentional substitution of a pipe supplying chlorine by one containing fuel (Cachia 2003). Increasing reliance on IT is envisioned as being key to avoiding these intrusions.

Implications of Growth Trends in the Water Supply and Wastewater Treatment Industries for IT

The dependency of water supply and wastewater treatment on IT is likely to grow as service areas expand in size. As indicated earlier such expansion generally results in more dispersed users and more concentrated or centralized providers of water infrastructure services, which increases the demand for IT-driven systems.

The increasing centralization of water infrastructure is borne out by the fact that fewer larger systems serve a much larger proportion of the population than are served by the more numerous smaller systems. Though exact estimates differ (Solley *et al.* 1998: 61; US EPA 2000), the number of wastewater treatment facilities in the US has been rising dramatically and serving more people as population has increased and as legislation has required connection to these facilities. In 1950, the number of POTWs was 11,784 serving 91.8 million people and by 1996 was 16,024 serving 189.7 million people (US EPA 2000: 2–32). Increasing centralization is indicated by the fact that in 1950, the average POTW served 7,790 people, whereas in 1996 the average population served was 11,838. Another way of looking at it is that between 1950 and 1996, the number of treatment plants increased by about 50 percent while the population served increased by 100 percent.

A similar pattern of concentration exists for water supply treatment systems. About a third of the community water systems in the US serve about 90 percent of the population (US EPA 2002: 10). This is exemplified by the fact that only 361 community water systems, each serving more than 100,000 people (or 6.8 percent of all community water supply systems), together serve an estimated 120 million people. This represents 45 percent of the population served by community systems, or over a third of the total US population.

Customer Needs

Infrastructure managers interface with the public primarily in the areas of customer accounts, marketing and response to preferences and complaints. IT has been drawn on more and more to provide these services, with mixed results.

Customer Preferences for Water Quality

Environmental surveys are used to identify the expressed beliefs of people toward the environment in general and services to ensure water quality in particular for customer satisfaction and marketing purposes. The general public has consistently ranked water quality as an area of concern, whether it is drinking water or the waterways into which wastewater is discharged. Experts and business leaders also rank the issue of water quality high. When these leaders were surveyed by the Water Environment Federation on their preferences for environmental indicators to include in a global assessment of environmental performance, they ranked water indicators relatively very high. Water pollution and water quality each ranked 3.9 and water quantity ranked 3.6 out of a possible score of 5 (most important), though 4.0 was the highest rank any of the indicators received (Esty and Cornelius 2002: 100).

Billing Systems

The ability to accurately track and assess costs of water treatment, whether for potable water supply or to maintain ambient water quality, ultimately falls to the billing and accounting systems. These systems have become increasingly automated through the use of IT, requiring new management skills. An example is the Automated Billing System in Sydney, Australia. In May of 2003, Sydney Water installed a new computer based billing system, the Customer Information and Billing System (CIBS) to provide proper accounting and billing services to its customers. According to Totaro (2003), serious management flaws resulted in the project costing over four times its original estimate, only to be ultimately abandoned. The fundamental problem appeared to be the inability of the management, including the Board, to interface effectively with the new technology or even understand it or the work of the contractors hired to create it. The problem of infrastructure organizations providing an effective interface with IT organizations is a pervasive one, given the fact that IT advances at such a rapid rate.

The Role of IT for Product and Process

Trends in Automation

The automation of water infrastructure has been increasing for decades and its benefits are being recognized to a greater degree (Hill *et al.* 2002). Moreover, digitization of one type of IT, sensors, is expected to increase in the future. Olsson and Newell (1999: 612) note that at the present time, the transmission of signals or information from sensors is primarily accomplished via electrical current, but in the future digital systems are expected to dominate.

Drivers of IT use for water infrastructure are related to increases in the number of requirements and the need for greater operational speed. Rosen, Larsson, Jeppsson, and Yuan (2002: 299) have attributed increased use of IT in water treatment to "a significant increase in the number of process variables that can be reliably measured on-line together with an increased knowledge of the biochemical processes." Increasing complexity has increased the speed system controllers have to act, and this in part drives automation. Beck and Lin (2003: 44) observe that the speed of system disturbances necessitates the need for automation of information gathering and control systems that respond to data in real-time, since the ability of human intervention is limited at such speeds.

Finally, surveys of the use of early warning devices reveal that IT is an important aspect of the performance of such devices (Love and Bott 2000).

A number of indicators have pointed to the fact that while usage of IT has increased, the results have been successful. Interest in automation of control has been around for over thirty years (Hill *et al.* 2002). Sensors have become a popular means of surveillance and control at many different points in water systems – both distribution and treatment. Yet, the prevalence of use is actually lower than expected given the level of interest. The conclusions of the survey by Hill *et al.* (2002: 1–6) are that "many wastewater treatment facilities utilize little automation. In those facilities utilizing automation, only about half of the control loops run in the automatic mode." In spite of the growing interest, Hill *et al.* (2002: ES-1) found that the use of automatic instrumentation and control systems are primarily restricted to usage for "simpler measurements such as flow, pressure, level, position, and temperature" that are relatively resistant to automation.

Uses of IT in Water and Wastewater Infrastructure

IT is used in a very wide range of water and wastewater management operations. According to Olsson and Newell (1999: 4–6), reasons for employing systematic controls include:

- building knowledge of the system for communication, transparency and accountability requirements of the public
- meeting regulatory requirements for quality
- identifying ways to achieve a high level of cost-effectiveness
- planning for changes in capacity
- coping with highly variable conditions (e.g. concentrations of pollutants, operating conditions, etc.)
- enhancing the ability to foresee and deal with unexpected loads or demands, known as "upset" conditions.

Three of these are discussed below: system knowledge for condition assessment, operations and maintenance; regulatory requirements; and early warning systems to detect upset conditions. The issue of communication with the public – both users and others impacted – cuts across each of these areas.

System Knowledge for Condition Assessment, Operations, and Maintenance

Surveillance and monitoring for system knowledge to more accurately assess the water supply resource base has been a traditional use of IT for water resources management (UNESCO 2003: 16). IT use for routine control for operations and maintenance came later, since many systems operated under mechanical or manual controls linked directly to power supplies without IT intermediaries. Before identifying some of the uses of IT for system knowledge, it is important to frame this issue in terms of measures of system condition. IT is inevitably seen as a key element in closing the very wide gap between the existing condition of water and wastewater infrastructure and the goals to improve condition.

The reliability and condition of water supply and wastewater infrastructure systems are routinely assessed in the US by professional organizations and government. The 2003 American Society of Civil Engineers (ASCE 2003) update of its infrastructure report card ranked the quality of water infrastructure nationwide relatively low, as a grade of D, with only two other areas of infrastructure at or below that estimate. They estimate an annual need (as a shortfall) of US$11 billion per year for water supply and US$12 billion a year shortfall for wastewater. The US EPA (2002) has conducted a comprehensive needs assessment for both water supply and wastewater infrastructure, and its estimate of the cost of closing the gap between what exists and what is needed in the US over 2000 through 2019 for clean water and water supply is shown in Table 5.1 (US EPA 2002: 43).

Gaps are in part driven by or at least exacerbated by aging infrastructure. Below are estimated design lifetimes of major components of water infrastructure. The EPA gap analysis took into account the extent to which existing facilities now exceed their estimated design lifetimes summarized in Table 5.2. What is particularly significant is the generally long lifetime of each of these components, especially relative to the time period of change of IT (not lifetime), which can be a year or less.

IT for Leak Detection

One maintenance area in which IT is heavily used is leak detection. Leakage is the percentage of total water lost in the distribution system through cracks,

Table 5.1 *Estimated Needs (in Billions of Current Dollars) for Drinking Water and Wastewater Treatment Infrastructure, US, 2000 – 2019 (assuming no revenue growth)*

	Minimum	*Maximum*
Clean water		
Capital	$331	$450
Operations and maintenance	72	229
Drinking water		
Capital	154	446
Operations and maintenance	0	445

Source: Summarized from US Environmental Protection Agency, Office of Water. *The Clean Water and Drinking Water Infrastructure Gap Analysis*, September 2002, EPA-816-R-02-020, p. 43.

Table 5.2 *Estimated Lifetimes of Water Supply and Wastewater Infrastructure (in years)*

Component	*Clean Water*	*Drinking Water*
Water/Wastewater collection	80–100 (collections)	50–80 (reservoirs, dams)
Treatment plants – concrete structures	50	60–70
Treatment plants – mechanical and electrical	15–25	15–25
Mains	25 (force mains)	65–95 (trunk mains)
Pumping stations – concrete structures	50	60–70
Pumping stations mechanical and electrical	15	25
Distribution	90–100 (interceptors)	65-95 (distribution mains)

Source: Summarized from US Environmental Protection Agency, Office of Water. *The Clean Water and Drinking Water Infrastructure Gap Analysis*. September 2002. EPA-816-R-02-020, pp. 11, Table 2.1.

breaks or poor connections once it has left the production facility, rather than in terms of water loss in storage systems, such as reservoirs, prior to entering the production facility. Many different terms are used to describe water lost from distribution systems, such as nonrevenue water (NRW) or "unaccounted for" water (Chowdhury *et al.* 2002: 64–5) and is usually defined for service areas. A typical or "acceptable" percentage is considered to be under six percent, though this varies for developed and developing countries (Chowdhury *et al.* 2002: 72, for example, citing (Chowdhury 1995) indicates that 20 percent is usually considered acceptable for developing countries and 10–15 percent for developed countries). Areas deviate from these ideals substantially across different cities as well as between developed and developing nations. For example, Chowdhury *et al.* (2002: 64) cite averages of 13 percent and 26 percent for these areas from Shipman (1978) and ranges from Farley (1991) from Germany's nine percent to Malaysia's 43 percent, with an average of between 20–30 percent. They note that within Bangladesh alone, the percentage of water supply that is NRW varies from 22–56 percent (Chowdhury *et al.* 2002: 64). More of these patterns and trends in water infrastructure in developing countries are described in Chapter 10.

The existence of leakages has considerable economic consequences for the producers and the users (as well as being wasteful), and in the extreme, can produce water insufficiency. Thus, computerized detection systems are used to detect leakage in real-time (usually through changes in water pressure or flow), the quantity of leakage, and deformities or other conditions in the distribution system that could be contributing to the problem. The interpretation of measurements is complicated by an opposite phenomenon – infiltration – where water enters cracks and breaks, rather than exiting the systems. Alternatives to computerized systems are visual inspection, however, this is difficult to accomplish for underground piping, with the exception of the atypical configuration of water supply pipes in human-accessible sewers in Paris, France.

IT TO MANAGE LONG WATER TRANSMISSION DISTANCES
A couple of cases illustrate how IT is being used to enable wastewater treatment plants to overcome large distances by monitoring facilities from remote locations.

Case: Austin, TX. Flow meters to measure overland flow entering water collection basins are often located in remote locations and are numerous, given the need to obtain data from different areas. According to the account by WaterWorld (2003a), the City of Austin, Texas attempted to use IT to overcome the considerable fragmentation and cumbersome nature of its flow monitoring apparatus, which was very labor intensive, requiring personnel to go into the field to check the meters. Field locations varied from very remote sites to those in downtown traffic areas. They encountered problems with lost meter readings due to construction damage and inclement weather. They switched from a data transmission system that depended on land lines to a wireless system. Although it initially required manual installation, the new system reduced the amount of manual field work as well as the power requirements.

Case: East Bay Municipal Utility District (MUD), CA. East Bay MUD provides water services to 1.3 million people and wastewater treatment services to 650,000 people in the San Francisco Bay area. It has employed a high frequency wireless broadband technology developed by Motorola to allow communication to occur over longer distances and to remote locations (WaterWorld 2003b). The 35-mile distance over which the technology monitors components of the wastewater treatment facility (e.g. pumps and water storage facilities) represents a record distance for that technology, and the plan is to eventually connect 200 locations. According to the WaterWorld account, the Motorola system, called the "Canopy" system, provides "high-speed Internet access utilizing the unlicensed UNII bands (5.25–5.35GHz or 5.725–5.825GHz)."

Regulatory Requirements for Quality
The federal Clean Water Act and equivalent state legislation prompted extensive monitoring to ensure the safety of natural waterways for fishing, swimming, potable water, and public health. The extensiveness of monitoring required for public health is reflected in the large number of waterborne diseases, which originate from either sewage or disease carrying vectors using stagnant water as a medium (UNESCO 2003: 103). An estimated 1.3 million waterborne disease cases occur in the US annually (Water Infrastructure Network 2000: 1–2, citing Payment *et al.* 1997). About half of diarrhoeal diseases are attributed to water (UNESCO 2003: 105; Prüss *et al.* 2002). Both water supply and wastewater treatment use monitoring systems for water quality that rely on information technologies to detect, record, and support the analysis of quality for public health.

Monitoring supports water quality-related activities other than public health, such as the 1.8 billion annual water dependent recreation trips and billions in economic value associated with water-related commerce and jobs that depend on water quality (Water Infrastructure Network 2000: 1–2). More demands are being placed on detection technology as standards have become more stringent for all water uses.

If one element of the regulatory monitoring system is deficient, the rest of the information system can suffer. For example, Seelye (2003: A1) points out that the US Environmental Protection Agency's computing capability for water quality monitoring has suffered from insufficient funds for fully automating the system for real-time, electronic recording of data, and facility information is still entered manually.

Early Warning Systems for Unexpected, Extreme Events

There is a growing need for IT to detect and control severe upset conditions, particularly in biological processes that drive water and wastewater treatment plants. These upsets have numerous origins such as biological, chemical, or physical (including weather) agents or technological or economic disruptions. Such conditions can severely compromise the reliability of these systems. Love and Bott (2000: 1–2) define an upset event as an observed deterioration of a treatment process, usually transient. The uncertainty in the nature and timing of these events will pose challenges for the IT systems being used for routine conditions.

One set of events involves biological agents, such as sudden outbreaks of new, emerging disease-causing pathogens. Another set is chemical; for example, the presence of toxic contaminants such as metals, grease, and salt in unusual concentrations in influent waters can destroy bacteria used in treatment processes. A third set of conditions pertains to natural hazards such as aberrant weather conditions, in particular flooding (treatment plants have to bypass water that exceeds three times their hydraulic capacity). A fourth set is technologically related to the processes, raw material supply chains, and facilities needed to sustain them, and include insufficient influent or quantity to sustain the processes, irregular power supplies or outages, explosions from unexpected combinations of substances or physical conditions, equipment malfunction, or operator errors. Grigg (2003: 83) identifies some of these and adds additional threats such as financial loss and loss of public confidence (resulting in panic or mass terror).

These events place unusual demands on IT and cause changes in the way that automated surveillance and control systems normally operate. Channels through which the information will have to pass are likely to be more extensive, no longer restricted to the local plant analysts and governmental oversight agencies. The timeframe for obtaining and delivering the information can increase. Quality controls needed for security may be very different and more extensive than those required under conventional quality assurance and quality control procedures because of security needs and precautions to avoid the further spread of a contaminant. The International Life Sciences Institute (ILSI) (ILSI 1999: 11) and others have identified additional requirements for early warning systems that rely upon IT. These include coverage of a wide range of threats, yet progress is still needed to identify serious threats, minimize false positives and negatives, enhance portability, operate in hostile environments, provide full spatial and temporal coverage (continuous use over 24/7 and 365 days a year), and provide compatibility (interoperability) with existing infrastructure.

Specific IT Applications for Water Systems

Objectives for the Introduction of IT

A major purpose for introducing IT into water infrastructure is to generate data to manage these systems better. Water quality data are critical to a wide range of management functions for both water supply and wastewater treatment. Data are the basis for proactive analysis and management of water infrastructure assets, system control, emergency management, vulnerability and damage assessment, security, customer communications, and regulatory compliance. Databases for these purposes range geographically from site-specific information to data for entire watersheds. Irrinki (2000) highlights the use of the Internet for various management functions such as selecting engineering firms, managing the development process from design to construction, overseeing operations and maintenance by importing data collected through Supervisory Control and Data Acquisition Systems (SCADA), connecting data to wireless systems (Irrinki 2000: 30–1), analyzing assets to optimize investments, using online billing and communicating with customers more effectively (Irrinki 2000: 32). Types of IT technologies used or potentially of use in water infrastructure and their advantages and disadvantages are discussed below with an emphasis on sensor technology.

Types of Technologies

Sensing technologies are at the interface between analytical protocols, such as bioassays, and the infrastructure technology. In water systems, sensors are very specialized, and some of the newer generic categories are biochips and optical sensors. Optical sensors, which involve light sensitivity to substances, were described in detail in Chapter 2, along with the advantages they offer in resisting environmental damage. Biochips and their application to water systems are described briefly below.

Biochips

Biochips are a series of technologies from molecular biology that emerged during the 1990s for the detection of genes that are now being used in water and wastewater treatment processes for diagnostic purposes. The terminology varies, and according to Shi (1998–2001) includes "biochip, DNA chip, DNA microarray, and gene array." Dozens of products and applications now exist all over the world. Applications include the water systems of the City of Paris and Atlanta, GA and a new product called the "Doodlebug" (Grow *et al.* 2003). Biochips are a special case of a sensor, and operate in the following way (Grow *et al.* 2003: 1–2):

> Different biomolecules that are specific for different targets are immobilized in a compact, two-dimensional array of pixels. The sample is introduced to the chip so that binding can take place between the targets in the sample and the biomolecules in the array: and a transducer then detects binding at each of the pixels. Pattern recognition software is used to identify the targets based on the pixels where binding is detected.

Although the binding of the molecule usually depends on what are called "labels" such as enzymes, some bypass that step and detect the targets themselves using light waves. Newer approaches avoid culturing cells prior to applying detection systems, which slows down the analysis and makes it difficult to differentiate a target organism from other organisms typically found in environmental samples (Higgins 2002: ES-1).

Sensors used in water systems are highly specialized, and a large number are required to cover the entire array of substances to be measured. For example, for water quality, turbidimeters or nephelometers measure turbidity, settlometers measure settleable solids, and luminometers measure purity or toxicity (Owen 2002). For flow, various devices exist such as Venturi meters or Parshall flume devices depending on whether flow is measured under pressure or not (Olsson and Newell 1999: 572–5). Dozens of companies are marketing equipment for automated digital sensing of water systems.

Supervisory Control and Data Acquisition (SCADA)

SCADA systems are a special application of IT-based sensor systems. Fitzgerald (2002) refers to them as a type of programmable logic controller (PLC).

SCADA systems used for the detection of water system condition, primarily for distribution systems, grew in popularity over the decade of the 1990s. A survey taken in 2002 indicated that SCADA systems have and are likely to continue to increase in popularity within the water and wastewater industry. In the survey, drinking water utilities ranked SCADA second out of a possible nine types of upgrades and wastewater utilities ranked them lower (7th out of 10) (WaterWorld 2002).

SCADA systems operate in conjunction with other IT. Ideally, Finnan (2002) argues, a SCADA system connected with monitoring devices such as "video cameras, motion detectors, contact switches, keypad entry devices and card readers" and process analyzers for water quality contamination could detect and react to threats to water systems, by shutting them down or isolating them. Interfaces among system components are gradually being worked out. For example, according to Finnan (2002), digital cameras address interconnectivity of closed circuit television (CCTV) with PCs and the Internet; programming has addressed bandwidth limits within SCADA.

Advantages and Disadvantages

Sensors used for water infrastructure have many advantages and disadvantages that need to be taken into consideration in the decision to introduce them into water infrastructure. Olsson and Newell (1999: 565) note in the context of wastewater treatment plants that "sensors are the weakest part of the chain in real-time process control." Reasons include the fact that traditional wastewater treatment design may not always include sensors upfront in the initial design process and regulations use laboratory methods rather than data that sensors would generate as a frame of reference (Olsson and Newell 1999: 565). Strengths and weaknesses of sensors vary with the wastewater component to which they are applied.

First, the large number of sensors currently required to cover all needs across a wastewater treatment process creates management problems. Sensor technology has to be adaptable to changes in the pollutants that have to be monitored, new levels of detection, etc. Second, false positives can arise if something obstructs the sensor that is not necessarily the substance being measured. Olsson and Newell (1999) cite constraints in using flow meters for liquids that aren't clear, and in fact only two of the sensor technologies they identify in their review can handle water with substances in it. Third, the accuracy of the measurements is a function of the ranges they are programmed to detect, and mistakes can occur where measurements out of range went undetected. Fourth, some technologies require readjustment or recalibration to prevent "drift" (Olsson and Newell 1999: 567; Hill *et al.* 2002: 1–2). Fifth, Hill *et al.* (2002: 1–2) note a number of other problems related to operations, such as the requirement for complete mixing of the material being measured in order to ensure accuracy and conditions that lead to malfunctioning of various parts of the sensor. Olsson and Newell (1999: 7) for example note that "Most sensors today do not allow long periods without extensive maintenance." Sixth, sensor devices have fundamental problems in their ability to operate in complex water environments in which the medium can interact physically or chemically with the sensors, thus impeding their performance. Seventh, sensors, like many pieces of equipment, rely on electric power and water quality monitoring equipment can be compromised by intermittent power outages. Eighth, problems occur with respect to the inability to access appropriate locations and obtain real-time information for quick response. Ninth, a continuing challenge is the capability to convert the vast amount of data produced by sensors into usable information to build a knowledge base for managing water supply and wastewater infrastructure. Beck and Lin (2003: 43) observe how we may have become "data-rich, yet information-poor." This is exacerbated by the fragmentation at the point at which information is collected, since rather than having one sensor meet all requirements, a suite of specialized sensors is usually used.

Gillette *et al.* (c2002: 2) have suggested that the extensiveness or pervasiveness of IT in general and SCADA systems in particular can introduce or increase interdependencies, that is, they "have linked infrastructures in new and complex ways and have created new vulnerabilities." For example, many different kinds of infrastructure now depend on IT for vital information on how to control their operations. The growing dependency on IT raises issues particularly for security.

Security issues are key potential problems for sensors. On the one hand, it has been argued that SCADA offers considerable opportunities for promoting security because of the increased knowledge they provide. On the other hand, as system controllers, they are potential points of vulnerability, since they potentially enable access by terrorists within the system and at remote locations (even alternative control stations designed to promote security). SCADA vulnerability increases, according to Fitzgerald (2002), when it is connected to remote locations via dedicated telephone lines, radio signals and Internet connections that can be accessed by others outside of the control of central control-stations: an "unmanned location such as a pumping station, a water tower or a finished water-storage area." The variability of SCADA systems contributes to the difficulty of securing them. As Fitzgerald (2002) observes, "no two SCADA systems are alike." Therefore, individualized diagnoses for points of intervention are necessary. SCADA systems can be designed to be more secure. Fitzgerald (2002) indicates that dial-in features can be secured using call-back features that require passwords. Finnan (2002) suggests having the capability of disconnecting SCADA computers from outside areas, and securing them with backup power and redundant computer networks.

As IT systems increase in sophistication, a number of these technical problems have been overcome. Human resources problems will continue to pose a challenge, such as the ability of workers trained in water treatment technology to keep their knowledge and skills in IT up to the changing needs. This is compounded by the fact that operators of many small systems are not full-time.

Conclusions

Water is a resource whose ownership and management was historically so highly fragmented that it became one of the earliest case areas for the study of efficient, effective, and equitable resource management. (McKean 1958) This fragmentation occurs anywhere between water storage to distribution systems, largely due to the inability to resolve conflicts over shared resources and competing uses and users of those resources. UNESCO (2003: 10), for example, has pointed out that worldwide "there are 263 river basins that are shared by two or more nations and that are home for roughly 40 percent of the global population..." As a result, within nations, the trend has been toward increasing centralized management; however a substantial regional division of water supply and wastewater services still exists. In the last few decades, IT has been introduced as a vehicle for integration, at least technologically.

The introduction of IT into water resources management has faced the problem of any system within a system. New technologies are usually introduced into water management systems as add-ons with integration gradually occurring as the water systems accommodate to the new technology. Ways to integrate IT into water systems, especially in a security conscious environment, have been suggested by Heller *et al.* (1999) who point to innovations that allow external data sharing and also protect proprietary information, systems oriented analyses that support life cycle and industrial ecology principles, and a focus on end user or business objectives initially during the design and selection process for an IT system. Unfortunately, given the rapidity with which IT changes relative to the rate at which water technologies change, it is difficult for water systems to easily accommodate IT as an integrated system. This is compounded by human resources issues such as the tradition of plant workers often being part time. As a consequence, manual systems are still a tradition, even for security, though they may be combined with IT: "At New York City's watersheds and reservoirs, for example, more than 200 guards will be posted by the end of the year, up from 45 in 1998, said Geoffrey Ryan, a spokesman for the Department of Environmental Protection. Reinforced gates, new fencing and public access restrictions have been put in place to secure perimeters..." (Baker 2002).

Nevertheless, the use of IT in water systems has proliferated and in fact, some applications such as biochips have been leaders in the IT field in areas other than water. The next step will be to continue to integrate these systems effectively with data analysis and the conveyance of information to meet regulatory and customer needs, and to provide the necessary human resources and management skills to support these activities.

Acknowledgments

The author wishes to acknowledge Mara Cusker, Assistant Research Scientist, who conducted literature searches for this chapter, and ICIS graduate research assistants Carlos Restrepo and Cary Hirschstein for their reviews of earlier drafts.

References

American Society of Civil Engineers (ASCE) (2003) "2003 Progress Report – an Update to the 2001 Report Card," Washington, D.C.: ASCE.

Baker, I. (2002) "9/11 One Year Later," *Newsday*, September 8, p. W18.

Beck, M.B. and Lin, Z. (2003) "Transforming data into information," *Water Science and Technology*, 47(2): 43–51.

Cachia, P. (2003) "Saboteur contaminates water supply at Ta' Kandja," di-ve news, November 10, Online. Available at: www.di-ve.com/dive/portal/portal.jhtml?id=114070&pid=null (accessed December 4, 2003).

Chowdhury, M.A.I. (1995) *Unaccounted Water Management Model in Developing Countries*, Master's thesis, Brussels, Belgium, Katholic University of Leuven and Vrije Universiteit.

Chowdhury, M.A.I., Ahmed, M.F. and Gaffar, M.A. (2002) "Management of Nonrevenue Water in Four Cities of Bangladesh," *Journal of the American Water Works Association*, 94(8): 64–75.

Cole, D., Saunders, D. and Guzzon, J. (2003) "Water shut off in Willcox after contamination suspected," *Sierra Vista Herald*, October 15, Online. Available at: www.svherald.com/articles/2003/10/15/news/news4.txt (accessed December 4, 2003).

CTV.ca News Staff (2003) "Lax security at Montreal water plant revealed," ctv.ca, November 14,

Online. Available at: www.ctv.ca/servlet/ArticleNews/story/CTVNews/1068763646446_3/?hub=TopStories# (accessed December 5, 2003).

Esty, D. and Cornelius, P.K. (eds) (2002) *Environmental Performance Measurement: The Global Report 2001–2002*, New York, NY: Oxford University Press.

Farley, M. (1991) "Leakage and Unaccounted-for-Water Studies," Hydrabad, India: Proceedings of the Seminar on Unaccounted Water Management.

Finnan, K. (2002) "Water Security: The Role of the SCADA System," *WaterWorld*, Online. Available at: ww.pennnet.com/articles/article_display.cfm?Section=ARCHI&Subsection=Display&ARTICLE_ID=139219&KEYWORD=SCADA (accessed January 9, 2004).

Fitzgerald, D. (2002) "Rethinking SCADA Security in the Age of Terrorism," *WaterWorld*, Online. Available at: ww.pennnet.com/articles/article_display.cfm?Section=ARCHI&Subsection=Display&ARTICLE_ID=154295&KEYWORD=SCADA (accessed January 9, 2004).

Gardner-Outlaw, T. and Engleman, R. (1997) *Sustaining Water; Easing Scarcity: A Second Update*, Washington, D.C.: Popular Action International.

Gillette, J., Fisher, R., Peerenboom, J. and Whitfield, R. (c2002) "Analyzing Water/Wastewater Infrastructure Interdependencies," Argonne, IL: Infrastructure Assurance Center, Argonne National Laboratory.

Gleick, P.H. (2000) *The World's Water, 2000–2001*, Washington, D.C.: Island Press.

Grigg, N.S. (2003) "Water Utility Security: Multiple Hazards and Multiple Barriers," *Journal of Infrastructure Systems*, Vol. 9, No. 2, June 1.

Grow, A.E., Deal, M.S., Thompson, P.A. and Wood, L.L. (2003) *Evaluation of the Doodlebug: A Biochip for Detecting Waterborne Pathogens*, Washington, D.C.: Water Environment Research Foundation.

Heller, M. Von Sacken, E.W. and Gerstberger, R.L. (1999) "Water utilities as integrated businesses," *Journal of the American Water Works Association*, 91(11): 72–83.

Higgins, J.A. (2002) *Overcoming Molecular Sample Processing Limitations: RNA/DNA Extractions Strategies*, Washington, D.C.: Water Environment Research Foundation.

Hill, R.D., Manross, R.C., Davidson, E.V., Palmer, T.M., Ross, M.C. and Nutt, S.G. (2002) *Sensing and Control Systems: A Review of Municipal and Industrial Experiences*, Washington, D.C.: Water Environment Research Foundation.

International Life Sciences Institute (ISLI) (1999) "Early Warning Monitoring to Detect Hazardous Events in Water Supplies," Washington, D.C.: ILSI, Online. Available at: www.ilsi.org/file/EWM.pdf (accessed January 15, 2004).

Irrinki, S. (2000) "The Digital Utility. Information technology enhances performance," *Water Environment & Technology*, December, pp. 29–33.

Khan, A.S., Swerdlow, D.L. and Juranek, D.D. (2001) "Precautions against Biological and Chemical Terrorism Directed at Food and Water Supplies," *Public Health Reports*, Vol. 116: 3–14.

Kinne, D. (2003) "Divers find dirt, vandalism in water tanks," theolympian.com, May 21, Online. Available at: www.theolympian.com/home/news/20030521/communities/10872.shtml (accessed December 5, 2003).

Lissau, R. (2003) "Neighborhoods want to go for lake water," www.dailyherald.com, December 4, Online, Available at: www.dailyherald.com/search/main_story.asp?intID=37959120 (accessed December 5, 2003).

Local6.com (January 14, 2003) "Plant Break-In Prompts Water Alert In Volusia," local6.com, Online. Available at: www.local6.com/news/1906837/detail.html (accessed December 4, 2003).

Love, N.G. and Bott, C.B. (2000) *A Review and Needs Survey of Upset Early Warning Devices*, Washington, D.C.: Water Environment Research Foundation.

McKean, Roland N. (1958) *Efficiency in Government through Systems Analysis*, New York: John Wiley & Sons.

National Research Council (NRC) (2002) *Making the Nation Safer: The Role of Science and Technology in Countering Terrorism*, Washington, D.C.: National Academy Press.
Olsson, G. and Newell, B. (1999) *Wastewater Treatment Systems. Modelling, Diagnosis and Control*, London, UK: IWA Publishing.
Owen, C. (2002) "Remote Sampling Technology: Proactive Management of Surface Water and Development of Comprehensive Data Sets for 'Early Warning' Applications," Online. Available at: www.apprisetech.com (accessed January 15, 2004).
Payment, P., Siemiatycke, J., Richardson, L., Renaud, G., Franco, E. and Prevost, M. (1997) "A Prospective Epidemiologic Study of the Gastrointestinal Health effects Due to the Consumption of Water," *International Journal of Environmental Health Research*, March, pp. 5–32.
Prüss, A., Kay, D., Fewtrell, L. and Bartram, J. (2002) "Estimating the Burden of Disease from Water, Sanitation and Hygiene at a Global Level," *Environmental Health Perspectives*, Vol. 110, No. 5, pp. 537–42.
Rosen, C., Larsson, M., Jeppsson, U. and Yuan, Z. (2002) "A framework for extreme-event control in wastewater treatment," *Water Science and Technology*, 45(4–5): 299–308.
Seelye, Katharine Q. (2003) "US Report Faults Efforts to Track Water Pollution," *New York Times*, May 27, p. A1.
Shi, L. (1998–2001) "DNA Microarray (Genome Chip) – Monitoring the Genome on a Chip," Online. Available at: www.gene-chips.com (accessed May 31, 2001).
Shipman, H.R. (1978) "Water Metering Practices," *Aqua*, 27(2): 1.
Solley, W.B., Pierce, R.R. and Perlman, H.A. (1998) *Estimated Use of Water in the United States in 1995*, Denver, CO: US Geological Survey.
Totaro, P. (2003) "Water bill fiasco sends $61m down drain," *Sydney Morning Herald*, May 2, Online. Available at: www.smh.com.au/articles/2003/05/01/1051382046556.html (accessed January 15, 2004).
UNESCO (United Nations Educational, Scientific and Cultural Organization (2003) *Water for People, Water for Life*, Paris, France: UNESCO Publishing, March.
US Bureau of the Census (2003a) Online. Available at: www.census.gov/cgi-bin/ipc/popclockw (accessed January 15, 2004).
US Bureau of the Census (2003b) Online. Available at: www.census.gov/population/projections/nation/summary/np-t1.pdf (accessed January 15, 2004).
US Environmental Protection Agency (EPA) (2000) *Progress in Water Quality: An Evaluation of the National Investment in Municipal Wastewater Treatment*, Washington, D.C.: US EPA. EPA-832-R-00-008, June.
US Environmental Protection Agency (EPA) (2002) *The Clean Water and Drinking Water Infrastructure Gap Analysis*, EPA-816-R-02-020, September.
Water Infrastructure Network (2000) *Clean Safe Water for the 21st Century*, Washington, D.C.: Water Infrastructure Network.
WaterWorld (2002) "Survey Results Highlight Infrastructure Need," *WaterWorld*, December, Online. Available at: ww.pennnet.com/articles/article_display.cfm?Section=ARCHI&Subsection=Display&ARTICLE_ID=164883&KEYWORD=SCADA (accessed January 15, 2004).
WaterWorld (2003a) "Digital Communication Improves Remote Monitoring," *Water World*, July, Online. Available at: www.pennnet.com/Articles/Article_Display.cfm?Section+CURRI&ARTICLE_ID=182579&VERSION_NUM=1&p=41&x=y (accessed January 15, 2004).
WaterWorld (2003b) "East Bay MUD deploys wireless broadband technology to improve security, monitor sites," *WaterWorld*, July 11, Online. Available at: ww.pennnet.com/articles/article_display.cfm?Section=ONART&Category=PROJE&PUBLICATION_ID=41&ARTICLE_ID=181797 (accessed January 15, 2004).
WHO/UNICEF (World Health Organization/United Nations Children's Fund) (2000) *Global*

Water Supply and Sanitation Assessment 2000 Report, Geneva, Switzerland: WHO/UNICEF.

Wong, A. (2002) "Friend or Foe?" *Water Environment & Technology*, pp. 26–9.

Woodtv.com (October 31, 2003) "Grand Rapids water pumping station broken into," woodtv.com, Online. Available at: www.woodtv.com/Global/story.asp?S=1505394&nav=0RceIqBF (accessed December 4, 2003).

Wright, R.T. and Nebel, B.J. (2002) *Environmental Science. Toward a Sustainable Future*, Upper Saddle River, NJ: Prentice Hall.

Yang, C.T. and Kao, J.J. (1996) "An expert system for selecting and sequencing waste-water treatment processes," *Water Science and Technology*, 34(3–4): 347–53.

CHAPTER SIX

Transportation

Christopher Cluett

Introduction

The surface transportation infrastructure can be characterized in terms of three main components: the *physical infrastructure* that is made up of roads and pavements, highway systems, bicycle and walking paths, bridges and tunnels, railroads, waterways, and their supporting sub-structures such as signage, street lighting, and traffic signals; the *vehicle infrastructure* that includes automobiles, buses, trains, and boats, along with their propulsion, fuel, and control systems; and the *organizational/institutional infrastructure* that is composed of both public and private components, including federal, state and local governments, and agencies that manage surface transportation and private companies that supply the various system components. In the past several decades the *digital infrastructure*, composed of sensors, controls, communication systems, computers, and data, has become an increasingly indispensable and integrated part of the physical, vehicle and institutional transportation infrastructure.

The links between these infrastructure components are such a natural association that it is already hard to imagine transportation systems without the digital overlay. Consider coordinated traffic signal systems; automated electronic toll collection; ramp metering at freeway entrances; electronic changeable roadside message signs; real-time traffic information on the radio, TV and the Internet; or multi-modal trip planning tools on the Internet as just some examples of the merging of the digital with the traditional to aid our surface transportation systems. The rapidly expanding capabilities of the digital infrastructure are both a cause of and a response to a remarkable paradigm shift in how the physical surface transportation infrastructure is being managed and operated today, and the effect it is having on travelers. This shift is from relatively autonomous systems with little shared information in which everyone is largely on their own, and where operators focus on building and maintaining physical structures, to a system where information sharing is pervasive, where travelers have options and a basis for making travel choices, and operators are focused on operating the system efficiently and safely to meet the needs of their customer. Arguably this shift would not be possible without the digital infrastructure, and at the least it is greatly facilitated by the sharing of digital information. It is not a coincidence that the metaphor of the information superhighway is derived from transportation. This merging of two different infrastructures into a tightly coupled hybrid form holds great promise, and it is the impact of the new information technologies on the traditional surface transportation infrastructure that is the subject of this chapter.

Transportation Infrastructure Under Stress

The backbone of our nation's surface transportation infrastructure is more than the 46,000 miles of the interstate highway system that was constructed beginning in the 1950s and connects major urban centers across the country. The interstate represents less than two percent of all roads and highways but it carries an estimated 43 percent of all vehicle-miles traveled and about 70 percent of commercial traffic (Federal Highway Administration 2002). Construction and maintenance of this extensive highway infrastructure, including roads, bridges, tunnels, signals, rights-of-way, and related infrastructure has been a core mission of the federal and state departments of transportation for four decades.

At the same time as the interstate highway system was being built, the nation's public transportation system was in a state of dramatic decline. The number of annual transit trips, which had peaked in 1946 at 23.4 billion, began a long, three-decade decline, just as the highway construction program was launched, to a low of 6.5 billion in 1972. Transit use has since experienced a turnaround, rising modestly in the decade of the 1970s, and settling in the past several years at a currently estimated 9.4 billion transit trips per year. While the network of interstate highways and local connecting arterials has dramatically expanded over the past half century, the number of automobiles and miles driven has grown at an even more rapid pace.

In fact, vehicle miles traveled increased more rapidly than either the construction of additional miles of road infrastructure or the rate of population growth might have suggested. More people were buying cars, and driving more miles, until the "supply" of the transportation infrastructure could no longer keep pace with the "demand" placed upon it. To make matters worse, the decline in public transportation reduced the use of the most prominent alternative to automobile travel. The inevitable result has been increased congestion, reduced mobility, and widespread frustration among travelers. A modest increase in the center line miles of roadway has been accompanied by the addition of lanes to existing roadways. While this has served to help mitigate congestion pressures somewhat, the potential benefits are limited by inadequate resources for road construction and urban land use constraints.

As shown in Table 6.1, over the past decade from 1990 to 2000 the total miles of urban and rural roadway increased 1.8 percent, while vehicle-miles traveled on those roadways increased 28.2 percent. During this period the total population of the US grew by 13.1 percent. Some of the factors contributing to the rapid rise in vehicle miles traveled (VMT) include the growth in one-person households, increased suburbanization and disbursed land use patterns, higher automobile ownership, and the aging of the population coupled with higher rates of older driving (Rosenbloom 1997). The excess of VMT over infrastructure growth has characterized the disjuncture between the development of additional surface transportation infrastructure and the usage of that infrastructure, at least since the 1960s. The net result has been congestion and stress, in terms of both system deterioration and the patience of travelers and system operators trying to accommodate reduced mobility and access.

Traffic congestion ranks high on the list of pressing public policy issues and citizen concerns in the US as well as in many countries around the world, costing countries billions of dollars annually. Congestion exacts high economic and social costs on society in terms of delay, opportunity costs, lost productivity, wasted fuel, environmental pollution, and quality of life effects. The Texas Transportation Institute (TTI) publishes a periodic report on urban mobility patterns in the US that quantifies some of these economic and societal impacts (Schrank and Lomax 2003):

Table 6.1 *Extent and Use of the US Surface Transportation System, 1990–2000*

Surface Transportation	*1990*		*2000*		*1990–2000*	
	Miles	*VMT*[a]	*Miles*	*VMT*[a]	*Miles*	*VMT*[a]
Highway total	3,866,926	2,144,362	3,936,229	2,749,803	1.8%	28.2%
Urban	744,644		852,241		14.4%	
Interstates	11,527		13,379		16.1%	
Arterials	212,549		240,441		13.1%	
Local	520,568		598,421		15.0%	
Rural	3,122,282		3,083,988		−1.2%	
Interstates	33,547		33,048		−1.5%	
Arterials	958,850		941,421		−1.8%	
Local	2,129,885		2,109,519		−1.0%	
Freight/Intercity rail total	143,758	413	122,250	539	−15.0%	30.5%
Class 1 rail	119,758		99,250		−17.1%	
Amtrak	24,000		23,000		−4.2%	
Transit rail total	5,966	3,242	7,601	4,081	27.4%	25.9%
Commuter rail	4,132		5,209		26.1%	
Heavy rail	1,351		1,558		15.3%	
Light rail	483		834		72.7%	

Source: US Department of Transportation, Bureau of Transportation Statistics (2002b). Figures compiled from Table 1.1 (System mileage within the United States) and Table 1.32 (US vehicle-miles).

Note
a Vehicle-miles traveled (millions of miles).

- The cost of traffic congestion nationwide in 2001 totaled almost US$70 billion, representing a cost of 3.5 billion hours of extra travel time and 5.7 billion gallons of fuel wasted while sitting in traffic.
- The average delay is 26 hours per person per year (2001, 75 metropolitan areas).
- The average rush hour trip takes 32 percent more time than the same trip taken during non-rush hour conditions.

The TTI mobility project has recently taken a closer look at several strategies being used by transportation agencies to enhance mobility. They have found substantial reductions in overall travel delays can be obtained by using ramp metering on freeway entrances, more rapidly detecting and clearing incidents, and coordinating traffic signals on arterials and local streets. These three example strategies depend heavily on the integration of the digital with the physical infrastructure to enhance mobility.

The magnitude of this problem is not, of course, confined to the US experience alone. Official estimates of annual economic losses caused by traffic congestion in Japan exceed US$100 billion annually (ITS America 2002b). European countries, facing the same problems, have been on the forefront in the global effort to find viable solutions. Many Asian countries experience some of the world's worst urban congestion due to rapidly rising automobile ownership and antiquated road infrastructures that are unable to accommodate changes in travel patterns from non-motorized to motorized vehicles. Bangkok and Dhaka readily come to mind – two capital cities so choked with cars, buses, trucks, and pedestrians that cross-town travel time is measured in hours, not minutes, and air pollution has become a major health concern.

A Transportation Research Board (TRB) study suggests that while "more people will be driving more miles annually, by 2020 there will be little fundamental change in the current network of highways" (Research Technology Coordinating Committee, FHWA 1997). Transportation professionals generally agree that there is no "silver bullet" solution to this surface transportation "problem;" rather, a multifaceted approach is required that includes a mix of demand and supply management multi-modal approaches, and changes in policies and social habits (Johnson 1993). Each of these strategies depends upon and is enhanced by the evolution of digital information technologies (IT) that are increasingly being brought to bear on surface transportation systems.

This discussion of a surface transportation system seeking technology interventions to help relieve mobility stresses and enhance efficiency is predicated on a system operating within bounds of normalcy. However, the events of September 11, 2001 and subsequent threats of terrorism directed toward a vulnerable transportation infrastructure have dramatically altered perceptions of those boundary assumptions. Congress created the Transportation Security Administration within the US DOT to assume responsibility for the security of all modes of transportation. Digital technologies necessarily play a key role in efforts to enhance surface transportation security through enhanced communication and data exchange. Increased attention is being directed to vulnerability and risk assessments, threat detection, deterrence and prevention, monitoring, human factors perspectives on security, emergency response planning, and training (National Research Council 2002). Accurate and rapid exchange of information is a central element in threat detection and prevention. Examples of current efforts to make our transportation infrastructure more secure include the application of non-intrusive shipping container inspection, and technology to detect chemical or radioactive substances in cargo traveling along our surface transportation network (World Shipping Council 2003).

Surface transportation challenges and problems range greatly in complexity, (Sussman 2000: 5–9) and the next section examines approaches to integrating the digital and physical infrastructures in the search for solutions.

The Digital Infrastructure in Transportation

To help address growing transportation problems, the US Congress passed the Intermodal Surface Transportation Efficiency Act (ISTEA) of 1991. That was followed by the Transportation Equity Act for the 21st Century (TEA-21) of 1998. ISTEA and TEA-21 signaled a shift from a focus on construction and maintenance of the roadway infrastructure to the operation of a multi-modal transportation system. These pieces of legislation grew out of a recognition that the country could no longer simply build its way out of growing congestion. As noted earlier, the costs of constructing new highway infrastructure are enormous, and land use and environmental constraints are daunting. The system had to be operated more efficiently, and ISTEA recognized the need for state and local transportation authorities to play a greater role, tailoring the infrastructure to the needs of local communities, and general and commercial travelers. TEA-21 builds on ISTEA with emphasis on transportation safety, environmental protection, and expanded research. Furthermore, the globalization of economies and the greatly increased interdependence among nations has placed additional emphasis on the importance of continuing to be able to move people, goods, services, and information efficiently and at competitive costs, both nationally and internationally.

ISTEA and TEA-21 have authorized and supported the Intelligent Transportation Systems (ITS) program.[1] It is the ITS program that is the primary mechanism by which Information Technology (IT) is being brought to bear on surface transportation (ITS America 2002a; Institute of Transportation Engineers 2000). The ITS program addresses urban and rural transportation challenges, general highway travel, commercial travel, and public transportation. Innovative applications of the digital infrastructure are being applied to the physical transportation infrastructure, including roads, rail, and transit systems, as well as to the vehicles themselves, including cars, trucks, buses, and rail. Many ITS applications tend to be implemented in urban areas first, in part for economic and political reasons but primarily in response to the intensity of urban traffic problems and the numbers of persons affected. The evidence shows that the larger the urban area the more severe the congestion problems (Schrank and Lomax 2003). Nevertheless, rural applications are becoming increasingly widespread, and some of the more advanced states have well integrated systems throughout their state. Therefore, this classification is somewhat artificial but useful in helping to illustrate the range and depth of IT applications across the country.

ITS in Metropolitan Areas

The central elements of the ITS program in urban areas are illustrated in Figure 6.1, showing progress made between 1997 and 2000 in deploying the digital infrastructure. The US Department of Transportation (US DOT), Federal Highway Administration's ITS Joint Program Office manages and conducts extensive research and evaluation on ITS deployments across the country. They have been surveying over 2,200 state and local transportation agencies since 1996 to monitor progress in deploying ITS technologies and evaluating system performance. The message in Figure 6.1 is that rapid progress is being made in the effort to integrate IT into various elements of the surface transportation infrastructure, but a lot more needs to be done, particularly in the area of gathering data on traffic conditions and disseminating information in real-time to the public so travelers can make better travel choices.

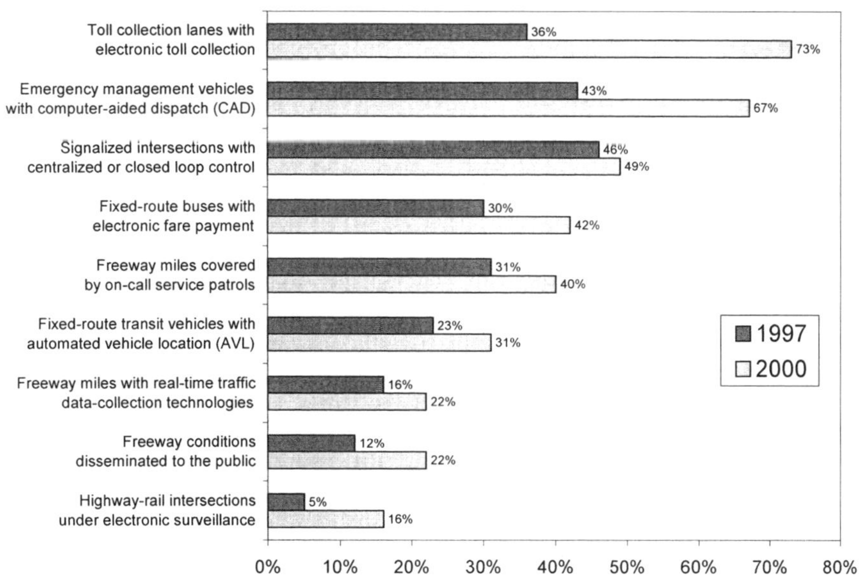

6.1 *ITS Infrastructure Deployment in 75 Metropolitan Areas, 1997 and 2000.*
Source: Adapted from US DOT, BTS 2002a; Figure 1, p. 40.

Digital infrastructure, or IT, that is being applied to surface transportation primarily in metropolitan areas includes the following infrastructure-based and vehicle-based components.

Infrastructure-Based IT

A variety of sensor technologies passively detect the presence of vehicles, or actively exchange information with vehicles that are equipped with transponders. Fiber optic cable is buried along miles of freeway right-of-way to provide rapid, high bandwidth exchange of data between sensors and cameras, and Traffic Management Centers (TMCs) where the data are processed and made available to both operators and the traveling public. Loop detectors embedded in the pavement, and sensors placed on signposts along the road side or overhead can detect and count vehicles, detect incidents (Adeli 2002), discriminate vehicle type and size, extract information from "intelligent" vehicles about the condition of the vehicle or the contents of commercial vehicles, weigh the vehicle while it is moving (weigh-in-motion), and verify credentials without requiring the vehicle to stop at a roadside station. Some of the digital technologies in use today include laser, microwave radar, ultrasonic, acoustic, and video imaging. Magnetic strips adhered to the pavement surface provide vehicle guidance in low visibility conditions. Sensors can detect emissions from passing vehicles to monitor environmental effects of transportation. A variety of wireless and Dedicated Short Range Communication (DSRC) devices offer the potential for greater automation of vehicles and road systems, along with safety improvements.

Linked via this communication system are systems to convey information back out to the highways to better inform motorists of changing conditions. US DOT (2000) and Maccubbin *et al.* (2003) provide current overviews of all the major ITS infrastructure components, along with an assessment of their benefits and costs. ITS highway systems include roadway message signs, known as Variable, Changeable, or Dynamic Message Signs (VMS, CMS, or DMS). They are connected by landline or wireless communications to a TMC. Standard message sets are digitally displayed on these signs to alert drivers to potential dangers on the roadway, such as restricted lanes, construction, accidents, traffic congestion, and weather. Messages can be posted and changed by central operators via telephone or other communication devices. VMS come in different sizes, depending on use and required length of message content, and they can be fixed or mobile, so that they could be moved, for example, to a construction site. VMS and CMS are finding many applications throughout the world, such as addressing congestion problems in Hong Kong and enhancing highway safety in California (Lam and Chan 2001; American City & County 2001).

Highway Advisory Radio (HAR) systems also offer travelers frequently updated information about what is happening on highway segments. HAR systems use DSRC, typically from either permanently mounted or mobile transmitters. They convey current traffic condition information over dedicated radio channels that are displayed to drivers on roadside signs. HAR is used for various types of traveler information, such as weather, road conditions, accidents, construction, tourist information, ferry schedules and queue status. Messages can be posted and adjusted by traffic operators by phone, wireless communications, or manual replacement of message tapes. As with VMS, mobile HARs allow state DOTs to use limited resources efficiently by moving them to the locations where they are needed most.

Closed-Circuit TV (CCTV) is being used extensively by state DOTs, for both metropolitan and rural applications, to display current road conditions to travelers and system operators either over the Internet or on standard television monitors. Communications can be landline (typically fiber optic to accommodate the high bandwidth video images) or wireless, and CCTV offers both snapshots and video clips. The cameras are often of the Pan-Tilt-Zoom variety that allow traffic centers to scan all segments of a roadway within the full field of view. They are useful for displaying traffic flow and congestion or focusing on accident scenes. The displays benefit drivers, system operators, maintenance personnel, and emergency service providers.

In July 2000 the Federal Communications Commission designated 511 as the nationwide three-digit telephone number for traveler information. Under the leadership of the American Association of State Highway and Transportation Officials (AASHTO) a 511 Deployment Coalition and coordination group have been established to guide state and local transportation agencies in setting up 511 programs across the country. As of November 2003, approximately 19 states or areas in states have operational 511 systems in place, covering about 17 percent of the population, and an estimated 35 jurisdictions are expected to have functioning 511 systems by the end of 2004. Travelers can call 511 to access the most current information on traffic and transit conditions and other related information that can assist them in better planning their trips.

Vehicle-Based IT

Sensors and transmitters in the vehicle can send information about the vehicle or its contents to external detectors along the roadway, to the driver of the vehicle, or to other vehicles. Information could include the presence of another vehicle in the blind spot (lateral collision avoidance), information about the vehicle in front or behind that could be used to regulate braking and acceleration (longitudinal collision avoidance), information about road hazards, night vision enhancement and head-up displays, and vehicle condition sensors that relay safety information to the driver, such as a tire failure warning system. Ranky (2002: 213) discusses the "increasingly software controlled digital car" that will be linked wirelessly to the household intranet network, warn the owner of parts needing replacement, and incorporate an aircraft-type black box function to remember the vehicle's internal processes in the event of a collision. Microsoft's vice-president of their automotive business unit expressed the software company's aspiration to "have one of our operating systems in every car on Earth," which he estimated to be 650 million vehicles currently, with a production rate of 50 million new vehicles every year (The Seattle Times 2003: C3). These kinds of systems are already showing up in high end car models, integrating many of the automobile's functions, providing navigation assistance, and employing hands-free voice-activated communications to avoid the safety risks of distracted driving.

The US DOT is evaluating many of these systems in its Intelligent Vehicle Initiative (IVI) program, though only limited results were published at the time of this writing. These include sensors in tanker trucks to detect the risk of roll-over, radar in snow plow trucks that allows safe operation in blinding snow storms (Estochen 2003), and short range side radars on metro buses to detect objects to the front and side of the vehicle. The goal of the IVI program is to increase safety and reduce the more than 41,000 deaths on US highways each year by aiding drivers in avoiding or coping with hazardous driving situations. A long-range vision for both infrastructure and vehicle-based sensors is a fully automated highway system (AHS) in which computer and communication systems assume responsibility for guidance, acceleration, and braking such that the driver need only be ready to take over when the system hands over control.

Sensors and signal devices can support capabilities such as signal pre-emption, allowing emergency vehicles or buses to alter traffic signal timing to facilitate efficient travel under certain authorized circumstances. Sensors on buses can count passengers, thereby facilitating the more efficient scheduling and management of a complex transit system. Digital card readers on buses offer regionalized fare cards that can consolidate multiple fare media and complex accounting systems across adjacent transit jurisdictions into a single, easy-to-use fare payment system. Coupled with on-line transit trip planners and real-time bus tracking software on the Internet, travelers will be able to more easily make multi-jurisdictional and multi-modal transit trips. Transit managers anticipate that such digital innovations can increase ridership and thereby reduce the burden on the rest of the surface transportation infrastructure. See Casey *et al.* (2000) for a comprehensive overview of a variety of technologies applied to public transportation systems.

Another common digital technology is finding many applications in support of the surface transportation infrastructure. The Global Positioning System (GPS) provides real-time location information displayed on map-based systems. GPS units placed on vehicles have led to systems that offer arrival-departure information for transit buses displayed at transfer stations, ferry locations displayed on the Internet (Washington State Department of Transportation 2003), assignment planning and tracking for snow plow operators (American City & County 2002), location and tracking of railroad cars or hazardous shipments, and many other related transportation uses. GPS is providing exceedingly accurate locational information, thereby facilitating a host of new transportation applications, including navigation and route guidance; the management of fleets of vehicles (Lahey 2003); emergency notification or mayday services; roadway maintenance; and intelligent vehicle applications. Expected benefits include increased safety, mobility, and operations efficiency. Automated Vehicle Location (AVL) systems are often GPS-enabled and are used, for example, for commercial vehicle and shipment tracking and a variety of transit applications (Strathman *et al.* 2002). Similarly, a transponder placed on a sealed shipping container can allow freight to be efficiently and securely tracked from loading for overseas shipment, travel into a US port of entry, transfer to a truck, and on through customs clearance at a border crossing, all without having to reopen the container for inspection.

In-Vehicle Navigation (IVN) systems typically depend on GPS and may be portable or permanently mounted in a vehicle. They are often coupled with Geographic Information Systems (GIS) and include up-to-date map databases to provide visual and/or auditory route guidance (turn-by-turn instructions) to the driver. They may include dynamic information on current road conditions and congestion so that a driver can be guided on alternate routes that circumvent problem areas.

Digital technology is installed on many commercial and emergency response vehicles. Examples included electronics on commercial trucks to record hours-of-service in place of manual entries into logbooks as required under existing regulations, thereby facilitating data entry by operators and data administration by fleet managers. Computer Aided Dispatch (CAD) systems allow highway patrol to examine license databases from their patrol vehicles or communicate details about an accident situation to emergency service providers.

As IT proliferates in various in-vehicle devices, the ability of the driver to use these systems effectively without distraction becomes a critical safety issue. Voice-activated and voice-recognition technology provide a way for the drivers to query the system and acquire needed information without taking their eyes off the road.

When a vehicle is involved in an accident, the driver may either not know the location (at least not with precision), or may be injured and not able to call for help. Automatic collision notification devices are automatically triggered by an impact and summon assistance without human intervention. They communicate exact location information, using GPS, to assist aid vehicles in quickly locating the accident site. Vehicles equipped with transmitting devices can serve as "probe" vehicles on the highway and automatically relay their speed and location to TMCs as another way to indicate where slow-downs may be occurring on the road system. Another potential probe technology involves the use of cell phones to anonymously track vehicles throughout the system and thereby calculate traffic speeds and flow patterns. This is an example of the use of the digital infrastructure that involves obvious public privacy concerns that need to be addressed.

Examples of both infrastructure-based and vehicle-based IT are summarized in Table 6.2.

ITS in Rural Areas

As is illustrated in Table 6.1, over 78 percent of the nation's roadway miles are in rural areas. The needs of rural surface transportation have both similarities and differences from urban areas, and therefore while the digital infrastructure overlaps urban and rural boundaries significantly, there are important differences in a number of respects. Rural roadways are more remote, are more difficult to maintain, experience different kinds of safety issues, are characterized by more variable terrain, and experience a wide range of weather extremes that, coupled with remoteness, can have particularly severe impacts on travel.

Table 6.2 *Examples of IT Used in Support of Surface Transportation*

Infrastructure-based IT	*Vehicle-based IT*
• Sensors detect and count vehicles • Fiber-optic cable for high speed communications • Weigh-in-motion for commercial vehicles • Magnetic lane guidance • Variable Message Signs (VMS) • Highway Advisory Radio (HAR) • Closed-Circuit TV (CCTV) and streaming video (with Internet display) • 511 traffic information national phone number • Traffic signal preemption for transit and emergency vehicles • Automated electronic toll collection • On-line Internet trip planning systems • Count-down real-time bus or train arrival digital signs • Ramp metering on freeways • Kiosks displaying static and real-time traffic and travel information • Road Weather Information Systems (RWIS)	• Transponders allow communication vehicle-to-vehicle and vehicle-to-infrastructure • Automated Vehicle Location (AVL) and Geographic Positioning System (GPS) • Forward and side-looking radars • Low visibility lane guidance • Head-up display • Truck roll-over warning system • Automated guidance, acceleration, and braking systems • Bus passenger counters and electronic fare card systems • In-vehicle navigation (IVN) • Computer Aided Dispatch for emergency vehicles • Voice activated and voice recognition systems • Automatic collision notification • Auto PC • Drowsy driver detection systems • Computerized vehicle propulsion systems

ITS in rural areas focuses on the following key areas:[2]

- emergency services
- tourism and travel information
- traffic management
- rural transit and mobility
- crash prevention and security
- operations and maintenance
- surface transportation and weather.

Providing effective emergency services in a remote rural area is difficult, both because it often takes a long time to detect that an emergency has occurred and because it takes time to respond. Obtaining useful traveler information in rural areas depends both on the ability to acquire relevant, current information and to effectively convey that to travelers. A high percentage of rural accidents are associated with drowsy drivers and run-off-the-road type crashes or, in certain parts of the country, impact with animals on the roadway (US DOT 1999). In-vehicle systems are being developed to detect drowsy driving and alert drivers in time to correct. Weather can be severe in many parts of the country where rural roadways predominate, and a variety of systems are being used to measure, interpret, forecast, and communicate conditions that may impact travel on a micro-climate scale.

Because of the remoteness and long distances involved, it is typically more cost-effective to use wireless communication strategies for a variety of rural transportation applications. A good example is a road-weather information system (RWIS) that collects and transmits road surface and sub-surface temperatures, wind speed and direction, precipitation, and video images. RWIS facilities are placed in areas particularly vulnerable to icing, winter storms, flooding, and related environmental conditions that can impact safe travel. These areas are often remote (such as a mountain pass). Wireless communications are used in both rural and urban settings with VMS, HAR, CCTV and other data gathering sensors that feed data streams into TMCs to facilitate traffic operations. Individual travelers also have access to new wireless technologies for sending and receiving information that is used for trip planning, summoning emergency services, and facilitating travel decision making.

In 1996, the North Dakota DOT initiated a 24-hour wireless call-in phone number – #SAFE (called Pound SAFE) – for travelers to use to get up-to-the-minute weather and traffic condition information across the state's roadways (Southerland 2001). This innovative application became an early model for the current 511 national traffic information phone number system. North Dakota's database that underlies this system is fed by data from over a dozen RWIS stations strategically located throughout the state. These stations are currently being augmented by deep subsurface probes that gather detailed temperature and moisture conditions below the pavement to allow DOT personnel to better assign weight restrictions to commercial vehicles during winter and spring freeze-thaw periods in order to minimize damage to road surfaces. The data from the RWIS provide detailed, advanced information as weather systems move across the state, and serve as an early warning system for road maintenance crews as well as for the traveling public seeking to plan their trips in advance.

Disasters, whether natural or man-made, often disproportionately impact rural areas. Furthermore, when events such as hurricanes, earthquakes, floods, hazardous materials spills, or national security emergencies prompt large population evacuations, the surface transportation infrastructure can be significantly affected. ITS is being applied in emergency situations to help respond to and recover from such events. The digital infrastructure is a critical component, providing information both about the disaster and about the surface transportation system at the disaster and on potential access routes. ITS guides the evacuating population away from disaster areas, and helps assign and track responders and emergency services in a coordinated way while facilitating communication among different agencies and coordinating through the regional and local TMCs to assure the effectiveness of the transportation system in the response and long-term recovery processes.

Issues in the Digital – Transportation Interface

The Transportation Research Board (TRB), a division of the National Research Council, periodically reviews the state of the nation's transportation system and prepares an overview of the important challenges facing users and providers of transportation services. The most recent in this report series identifies 14 critical issues (TRB 2001b), several of which are particularly relevant to this discussion.

- Traditional institutional and organizational structures that characterize the "patchwork of loosely connected modes encompassing 35,000 government owners of roads; tens of thousands of private carriers; and thousands of independent authorities responsible for ports, transit systems, and airports" (TRB 2001b: 5) are poorly integrated and ill-suited to effectively manage the future development of the transportation infrastructure. Many of these organizations are firmly rooted in a culture "created to build systems, not to manage operations." Some transportation agencies, operating in the public spotlight and subject to an uncertain funding base, tend to be conservative and reluctant to take undue risks with new technologies, thereby slowing the diffusion of the digital infrastructure in the transportation sector. However, others are leaders in deploying innovative ITS solutions to meet the needs of these infrastructures.

- Transportation organizations lack the skills, training, and often the vision to deal with the demands and opportunities provided by the emerging digital technologies, and educational institutions are facing the need to train the "New Transportation Professional" to deal with these complexities (Sussman 1999: 15). The TRB report notes the difficulties these agencies face in competing for and attracting skilled personnel. Outsourcing components of transportation agencies' IT function has emerged as one response to this problem, but that comes with real costs; agencies are finding they have to give up a measure of control over their IT functions, and quality assurance becomes an important issue.

- The TRB report asserts that "the merging of telecommunications and information technologies with transportation offers the greatest potential for innovation in transportation" (TRB 2001b: 8). They note, however, that the anticipated benefits associated with smoother traffic flows, ease of telecommuting, wireless access to traveler information, and substantial safety enhancements are potentially offset by societal concerns such as system reliability, privacy, liability, and driver distraction. Like any other innovation, the new digital infrastructure applied to surface transportation is certain to have predictable as well as unanticipated consequences.

At the foundation of any transportation system that is to be informed by the digital infrastructure are the data that flow through the arteries and veins of the infrastructure, providing vital system status information to operators and travelers. A central source of these data derives from sensors and systems to monitor such key attributes as traffic flow, incidents, and weather in real-time. Such roadway data are part of the ITS INFOstructure (Tarnoff 2002), operating at both a local and national level. Tarnoff reports projections from the US DOT Joint Program Office that indicate between 44 percent and 55 percent of urban freeways are likely to be instrumented by 2010. He points out that this is much less than desired, particularly in view of the fact that only six percent of urban arterials and no rural roadways have been instrumented. There remains a serious data gap that must be overcome before the real potential of the digital infrastructure can be realized in transportation.

Equally important as filling the data gap is the need to effectively integrate transportation data and information across a multi-modal network. Three key "building blocks" are posited for this latest transportation initiative, the Integrated Network of Transportation Information (INTI) (Taylor 2003). The first is gathering more and higher quality data, as partially addressed in the INFOstructure initiative. The second component is sharing data with users. This depends both on adherence to the ITS architecture and standards that support the ability of different systems to talk to each other, and on institutional arrangements that can overcome roadblocks to sharing information among different jurisdictions and agencies. The third INTI building block is actually putting the data to work effectively, and creating partnerships between the public and private sectors to accomplish this objective.

A decade ago, the Clinton Administration promoted the creation of a National Information Infrastructure (NII) to encourage and facilitate a broad exchange of data and information in all sectors of the nation's economy. ITS was the core part of the NII focused on the transportation sector. The Harvard Information Infrastructure Project, established in 1989 to explore issues and policies associated with a NII, characterized the information infrastructure in terms of the physical elements such as communication lines and connections, the knowledge infrastructure or information conveyed through this system, and the integration infrastructure or "systems and processes that bring together heterogeneous networks, computers, databases, and applications" (Branscomb and Keller 1996: 3). These three components were identified as critical aspects for implementing an ITS program.

In January 1996, the Secretary of US DOT announced Operation TimeSaver, along with a challenge to deploy a comprehensive Intelligent Transportation Infrastructure (ITI) in each of the 75 largest metropolitan areas. This ambitious program called for a model deployment of integrated ITS in a few selected metropolitan areas to serve as a showcase, outreach to build institutional awareness, and capacity building to train ITS professionals (US DOT 1996). Subsequent to and building on the earlier NII and the ITI initiatives, success with the INFOstructure and INTI initiatives holds great promise for the benefits that can be derived from the integration of the digital and physical transportation infrastructures. However, state DOTs are acutely aware of the challenges associated with rapid and massive increases in the amount of data that they may be expected to manage. Questions are raised about the costs of acquiring and processing the data, determining which data are most important, data quality assurance, the impact on decision making of vastly more data needing to be assimilated and understood, data retention policies, who needs access to various data, data storage and retrieval, and how the information can be controlled and security maintained.

The latest in this series of efforts to further develop the potential for integrating the transportation and digital infrastructures is called the Vehicle-Infrastructure Integration (VII) initiative, and it is just in the early stages of being defined. The VII seeks a nationwide capability for vehicles to communicate with their surrounding infrastructure and with each other. A major challenge for this initiative, as with the others that preceded it, is to find a viable business model that will attract the private sector into successful partnerships with the public sector to implement these systems.

Conclusions

This chapter began by describing the decline in mobility, causing significant societal and economic consequences, and noting that there is no silver bullet that can solve the nation's growing transportation problems. Rather, the consensus in the transportation industry is that a variety of approaches are needed to solve these problems, and each agency and part of the country will seek a strategy mix that best fits their needs and situation.

TTI points out that "there are many programs and strategies that may not significantly change the average mobility levels, but can reduce travel time variations and frustration with transportation services" (Schrank and Lomax 2003: v). Research on the factors that affect traveler satisfaction shows that travel time reliability is more important than periodic short travel times with high uncertainty, and foreknowledge of traffic conditions offers a "serenity" value that offsets the stresses of traffic delays.

The digital revolution is proceeding at a rapid pace, and the transportation community is scrambling to anticipate and take advantage of what the evolving technology offers. Western and Ran (2000: 6) point out that "as we move into the 21st century, we need to identify the areas of transportation in which IT innovations might develop; we then can begin to solve problems and meet challenges proactively." This is a central theme that emerged from the TRB's year 2000 field visit program: "...state DOTs and other transportation organizations are in a transition from trying to understand and cope with change to managing change"

(TRB 2001a: 1). Lakshmanan *et al.* (1997: 43) also echo this theme when they suggest that "we should think of the emerging transportation system at the state level in a dynamic, not in a static, context." They suggest that the flow of information facilitated by the digital infrastructure supports and encourages decision making at three levels: *strategic* with regard to future investments; *managerial* with regard to appropriate uses and handling of information; and *operational* with regard to using new data applied to better operating transportation assets. According to a national transportation working group, "information is central to the cooperative and informed decision-making process underlying [transportation] Asset Management," and information management has been identified as their number one policy priority (Infrastructure Renewal Working Group 2001: 9). The former Executive Director of the Federal Highway Administration anticipates that explosive growth in IT and e-business will be major drivers of change shaping the future course of transportation (Kane 2000: 6).

The organizational and institutional infrastructure components of surface transportation noted at the outset of this chapter are affected no less by the digital infrastructure than the physical and vehicle infrastructures. State Departments of Transportation are increasingly adopting e-government strategies (i.e. business transactions conducted over the Internet) for procurement of goods and services, for communicating with their customers, for managing construction bids for roads and bridges, and for conducting administrative functions such as payroll and travel. The digital infrastructure is imposing increased standardization of organizational processes in the surface transportation industry, such as in financial management and project management, just as it is requiring standardization in the hardware and software systems that support the roadside and vehicle systems. It has been argued that because digital information "can organize and manipulate other digital information, [it] creates its own infrastructure" (Kahin 1993: 138).

Along with the infusion of the Internet into government business, e-commerce has emerged as a significant economic force with the potential to "change the demand for transportation" according to a recent workshop sponsored by the Oak Ridge National Laboratory's Center for Transportation Analysis and the TRB.[3] This is due to the substitution of the electronic exchange of information in place of travel or goods shipment, and examples include electronic banking and shopping. At the individual level, telecommuting has the potential to reduce congestion and VMT by substituting working at home, or a satellite office away from the work place and closer to home (Special Issue 1993), for the usual journey to work. A nationwide time series analysis of the impacts of telecommuting on VMT has found a small but significant VMT reduction due to telecommuting of less than two percent of observed VMT (Choo *et al.* 2002). However, these findings do not account for countervailing effects of non-work trips, residential relocation, or latent demand for travel that are known to offset this beneficial short-term impact of telecommuting. Nevertheless, it is clear that the digital infrastructure is exerting profound and very complex forces on the transportation infrastructure and travel patterns that are only partially understood at this time.

The notion of relying on almost instantaneous, real-time data, both for system operations and individual trip-making decisions is a relatively new concept only recently made possible by the digital infrastructure. The compression of time scales involved will require major adjustments and new analytical tools (Coogan 2002: 16) to effectively integrate new information technologies, traditional planning processes, and a wide range of complex circumstances and needs faced by urban and rural transportation systems across the country. In the transportation field it is harder now to think in terms of two separate infrastructures, one digital and the other physical. As this chapter points out, the two have become tightly entwined. IT serves as an "enabling technology," a powerful integrating force for transportation systems that provides "tools and a framework to redesign [transportation's] internal infrastructure" (Kahin 1992: 137).

The continued integration of the digital and physical infrastructures and the successful adaptation of transportation agencies to a new hybrid infrastructure will require significant institutional and organizational adjustments. These will include adopting an organizational culture that is comfortable with the infusion of high technology, a more rapid pace of decision making, and a greater degree of customer orientation and responsiveness. Institutionally they will require an increased openness and willingness to share information within and among agencies and across jurisdictions, with more attention to joint problem-solving. In addition, in order to meet the needs of this hybrid infrastructure, new organizational structures are likely to be required that bridge across old boundaries and bring under one roof a wide variety of skills and experience to effectively manage the transportation infrastructure of tomorrow.

Notes

1 The USDOT defines ITS as follows: "Intelligent Transportation Systems represent the next step in the evolution of the nation's entire transportation system. As information technologies and advances in electronics continue to revolutionize all aspects of our modern-day world, from our homes and offices to our schools and even our recreation, they are also being applied to our transportation network. These technologies include the latest in computers, electronics, communications and safety systems" (see "FAQs: What's ITS" under http://www.its.dot.gov).

2 The Federal Highway Administration's Joint Program Office has developed a "tool box" of rural ITS projects and statewide applications from across the nation. http://www.itsdocs.fhwa.dot.gov//JPODOCS/REPTS_TE/13477.html.

3 A workshop on the "Impacts of the New Digital Economy on Transportation: Developing Research and Data Needs" was held September 14–15, 2000 in Washington, D.C. (see "E-Commerce's Impacts on Transportation" under http://web.ornl.gov/info/ornlreview/v33_3_00/ecommerce.htm).

References

Adeli, H. (2002) "Automatic Detection of Traffic Incidents Using Data Obtained From Sensors Embedded in Intelligent Freeways," *Sensor Review*, 22(2): 145–9.

American City & County (2001) *Message Signs Reduce Road's Collision Rates*, Issues and Trends, August, pp. 16–17.

American City & County (2002) *It's Snowing: Do you Know where Your Vehicles Are?* September, pp. 57–8.

Branscomb, L.M. and Keller, J.H. (eds) (1996) *Converging Infrastructures*, Cambridge, MA: The MIT Press.

Casey, R.F., Labell, L.N., Moniz, L., Royal, J.W., Sheehan, M., Sheehan, T., Brown, A., Foy, M., Zirker, M., Schweiger, C.L., Marks, B., Kaplan, B. and Parker, D. (2000) *Advanced Public Transportation Systems: The State of the Art. Update 2000*, DOT-VNTSC-FTA-99-5, US Department of Transportation.

Choo, S., Mokhtarian, P. and Salomon, I. (2002) *Impacts of Home-Based Telecommuting on Vehicle-Miles Traveled: A Nationwide Time Series Analysis*, ITS-Davis, Prepared for the California Energy Commission, Publication No. UCD-ITS-RR-02-05.

Coogan, M.A. (2002) "Models of Mobility: the outcomes supported by the data we collect," White paper presented at the Transportation Research Board Workshop on The Roadway INFOstructure, Irvine, CA, August.

Estochen, B. (2003) "Specialty Vehicle Platform Results for Intelligent Vehicle Initiative: Minnesota Field Operational Test," in Transportation Research Board, Transportation Research Record, *Intelligent Transportation Systems and Vehicle-Highway Automation 2003*, R1826.

Federal Highway Administration (2002) *Fiscal Year 2003 Performance Plan*, Online. Available at: www.fhwa.dot.gov/reports/2003plan/index.htm#preface3 (accessed October 2, 2003).

Infrastructure Renewal Working Group (2001) *Infrastructure Renewal Research Agenda*, Washington, D.C.: National R&T Partnership Forum.

Institute of Transportation Engineers (ed.) (2000) *Intelligent Transportation Primer*, Washington, D.C.: Author.

ITS America (2002a) *National Intelligent Transportation Systems Program Plan: a ten-year vision*, Washington, D.C.: Author.

ITS America (2002b) *ITS Fact-Finding Mission to Japan in Conjunction with ITS Asia Pacific Forum* Online. Available at: www.itsa.org/ITSNEWS.NSF/0/d9d3889ca9c6a31f85256b3a00798de1?OpenDocument (accessed October 2, 2003).

Johnson, E.W. (1993) *Avoiding the Collision of Cities and Cars: urban transportation policy for the twenty-first century*, Cambridge, MA: The American Academy of Arts and Science.

Kahin, B. (1993) "Information Technology and Information Infrastructure," in Branscomb, L. (ed.) *Empowering Technology*, Cambridge, MA: The MIT Press.

Kahin, B. (ed.) (1992) *Building Information Infrastructure*, New York: McGraw-Hill/Primis.

Kane, A.R. (2000). "Transportation in the New Millennium," *Transportation Quarterly*, 54(3): 5–9.

Lahey, L. (2003) "Eyes in the Sky Track Street Fleet," *Computing Canada*, 15, July 18.

Lakshmanan, T.R., Skinner, R.E. Jr. and Francois, F. (1997) "Role of Information in Transportation Data Content, Analysis, and Institutions," *Conference Proceedings 14: Information Needs to Support State and Local Transportation Decision Making Into the 21st Century*, pp. 42–52.

Lam, W.H.K. and Chan, K.S. (2001) "A Model for Assessing the Effects of Dynamic Travel Time Information Via Variable Message Signs," *Transportation*, Netherlands: Kluwer Academic Publishers, 28: 79–99.

Maccubbin, R.P., Staples, B.L. and Mercer, M.R. (2003) *Intelligent Transportation Systems Benefits and Costs: 2003 Update*, Report prepared by Mitretek Systems for the US Department of Transportation, ITS Joint Program Office, Report No. FHWA-OP-03-075, May.

National Research Council (2002) "Transportation Systems," in *Making the Nation Safer: The Role of Science and Technology in Countering Terrorism*, Washington, D.C.: National Academies Press.

Ranky, P.G. (2002) "Advanced Digital Automobile Sensor Applications," *Sensor Review*, 22(3), 213–17.

Research Technology Coordinating Committee (Federal Highway Administration) (1997) *The Future Highway Transportation System and Society: suggested research on impacts and interactions*, Washington, D.C.: Transportation Research Board, National Academy Press.

Rosenbloom, S. (1997) "Future Travel Demands: Impact of Current and Future Societal Trends," Appendix A in National Research Council, *The Future Highway Transportation System and Society*, Washington, D.C.: National Academy Press.

Schrank, D. and Lomax, T. (2003) *The 2003 Annual Urban Mobility Report*, College Station, TX; Texas Transportation Institute.

The Seattle Times (2003) *Microsoft Sees Vast Market: Every Car*, Report on the Technology, Tolls and Transportation Conference, the Discovery Institute's Cascadia Project, November 2003, Seattle, Washington, C3.

Southerland, R. (2001) "RWIS Eliminates Guesswork." *AC&C Product Review*, April/May, 4.

Special Issue (1993) "Part 3: The Information Age and the Future of Transportation: Mobility and Transport Services of the Information City," *Wheel Extended*, 16–17.

Strathman, J.G., Kimpel, T.J., Dueker, K.J., Gerhart, R.L. and Callas, S. (2002) "Evaluation of Transit Operations: Data Applications of Tri-Met's Automated Bus Dispatching System," *Transportation*, 29: 321–45.

Sussman, J. (1999) "The New Transportation Faculty: The Evolution to Engineering Systems," *Transportation Quarterly*, 53(3): 15–26.

Sussman, J. (2000) *Introduction to Transportation Systems*, Boston: Artech House.

Tarnoff, P.J. (2002) "Getting to the INFOstructure," White paper presented at the Transportation Research Board Workshop on The Roadway INFOstructure, Irvine, CA, August 2002.

Taylor, R. (2003) "Integrated Network of Transportation Information (INTI) Update," Paper presented at the ITS America Annual Meeting on ITS Industry Gathers to Chart a Course for the INTI, Minneapolis, MN, May 2003.

Transportation Research Board (2001a) "Managing Change in Transportation: the TRB's 2000 field visit program," *TR News 212*, Washington, D.C.: National Academy Press.

Transportation Research Board (2001b) "Critical Issues in Transportation 2002," *TR News 217* Washington, D.C.: National Academy Press.

US Department of Energy, Oak Ridge National Laboratory Review at: http://web.ornl.gov/info/ornlreview/v33_3_00/ecommerce.htm (accessed December 9, 2003).

US Department of Transportation, Bureau of Transportation Statistics (2002a) *Transportation Statistics Annual Report 2001* (BTS02-07) Washington, D.C.: US Government Printing Office.

US Department of Transportation, Bureau of Transportation Statistics (2002b) *National Transportation Statistics 2002* (BTS02-08) Washington, D.C.: US Government Printing Office.

US Department of Transportation, Federal Highway Administration (1996) *Building the ITI: Putting the National Architecture into Action*, FHWA-JPO-96-011, HVH-1/4-96(10M)QE, Washington, D.C.: US Government Printing Office.

US Department of Transportation, Federal Highway Administration (2000) *What Have We Learned About Intelligent Transportation Systems?* FHWA-OP-01-006; EDL# 13316 December.

US Department of Transportation, Intelligent Transportation Systems at: http://www.its.dot.gov/faqs.htm#What%20is%20ITS? (accessed December 9, 2003).

US Department of Transportation, National Highway Traffic Safety Administration (1999) *Run-Off-Road Collision Avoidance Using IVHS Countermeasures*, Final Report (DOT HS 809 170) December.

US DOT's Rural Intelligent Transportation Systems (ITS) Program at: http://www.itsdocs.fhwa.dot.gov//JPODOCS/REPTS_TE/13477.html (accessed December 9, 2003).

Washington State Department of Transportation, Washington State Ferries (2003) *Vessel Watch.* On-line ferry tracking at: http://www.wsdot.wa.gov/ferries/commuter_updates/vesselwatch/.
Western, J.L. and Ran, B. (2000) "Information Technology in Transportation: key issues and a look forward," *Transportation in the New Millennium*, Washington, D.C.: Transportation Research Board, National Academy Press.
World Shipping Council (2003) *In-Transit Container Security Enhancement*, September 9.

CHAPTER SEVEN

Electricity

Massoud Amin

Introduction: Real-time Systems

The massive August 14, 2003 power outage in the United States and Canada evoked eerie reminders of what shook our world on September 11, 2001. While early reports indicated no evidence of terrorism in this outage or those in the UK and Italy in 2003, the cascading blackouts underscored our electricity infrastructure's vulnerabilities (Amin 2003). This infrastructure affects us all. Are we prepared for future episodes? What role does IT play in predicting, averting and/or managing such crises?

Our economy places increased demand for reliable, disturbance-free electricity. The electric power grid is quite different from other infrastructure systems such as gas, oil or water networks. A distinguishing characteristic of electricity, for example, is that there is no way to store significant amounts of energy; thus the system is fundamentally operating in real-time. For this and related reasons, energy infrastructure systems have a unique combination of characteristics that makes control and reliable operation challenging:

- attacks and disturbances can lead to widespread failure almost instantaneously

- billions of distributed heterogeneous infrastructure components are tightly interconnected

- a variety of participants – owners, operators, sellers, buyers, customers, data and information providers, data and information users – interact at many points

- the number of possible interactions increases dramatically as participants are added – no single centralized entity can evaluate, monitor, and manage them in real time

- the relationships and interdependencies are too complex for conventional mathematical theories and control methods.

These characteristics create unique challenges in modeling, prediction, simulation, cause and effect relationships, analysis, optimization, and control, which have important implications for the use of IT for electric power. This chapter addresses these challenges by first presenting the technologies involved in the electricity infrastructure and then considering management and policy challenges to the effective performance both in the short and long term.

Technology

The North American Power System Example

The North American power network may realistically be considered to be the largest and most complex machine in the world – its transmission lines connect all the electric generation and distribution on the continent. In that respect, it exemplifies many of the complexities of electric power infrastructure and how IT can address them. This network represents an enormous investment, including over 15,000 generators in 10,000 power plants, and hundreds of thousands of miles of transmission lines and distribution networks whose estimated worth is over US$800 billion. In 2000, transmission and distribution was valued at US$358 billion (EIA 2003; EPRI 1999–2003).

At its most fundamental level, the network's transmission lines form a vertically integrated hierarchical network consisting of the generation layer (noted above) and then three network levels (Kundur 1994). The first level is the *transmission* network, which is meshed networks combining extra-high voltage (above 300 kV) and high voltage (100–300 kV), connected to large generation units and very large customers and, via tie-lines, to neighboring transmission networks and to the sub-transmission level. The second level is *sub-transmission*, which consists of a radial or weakly coupled network including some high voltage (100–300 kV) but typically 5–15 kV, connected to large customers and medium-size generators. Finally, the third network level is *distribution*, which is typically a tree network including low voltage (110–15 or 220–40 V) and medium voltage (1–100 kV) connected to small generators, medium-size customers, and local low-voltage networks for small customers.

In its adaptation to disturbances, a power system can be characterized as having multiple states, or "modes," during which specific operational and control actions and reactions take place: normal, disturbance, and restorative. In the normal mode, the priority is on economic dispatch, load frequency control, maintenance, and forecasting. In the disturbance mode, attention shifts to faults, instability, and load shedding. In the restorative mode, priorities include re-scheduling, re-synchronization, and load restoration. Some authors include an Alert Mode before a disturbance actually affects the system (Dy Liacco 1967). Others add a System Failure Mode before restoration is attempted (Fink 1978).

Temporal and Operational Attributes of Electric Power Systems

As noted above, a distinguishing characteristic of electricity systems is its real-time instantaneous nature; this characteristic dramatically affects the time-period for engaging in normal, disturbance and restorative management activities. As noted in the table below, the majority of these activities take place in an extraordinarily short period of time.

Yet in spite of the electrical infrastructure's importance and complexity, this network has evolved without formal analysis of its system-wide implications, including diminished transmission and generation shock-absorber capacity under the forces of deregulation, the digital economy, and interaction with other infrastructures. This chapter provides a starting point for such an analysis, beginning first with technological issues surrounding the network's reliability and operational management and then moving to consider critical management and policy issues.

Reliability Issues

Several cascading failures during the past 40 years spotlighted our need to understand the phenomena associated with power systems, and development of emergency controls and restoration. Widespread outages and recent huge price spikes raised public concern about national grid reliability (DOE; EPRI 1999–2003;

Table 7.1 *Time Hierarchy of Power Systems*

Action/Operation	*Time Frame*
Wave effects (fast dynamics, lightning caused overvoltages)	Microseconds to milliseconds
Switching overvoltages	Milliseconds
Fault protection	100 milliseconds or a few cycles
Electromagnetic effects in machine windings	Milliseconds to seconds
Stability	60 cycles or 1 second
Stability Augmentation	Seconds
Electromechanical effects of oscillations in motors and generators	Milliseconds to minutes
Tie line load frequency control	1 to 10 seconds; ongoing
Economic load dispatch	10 seconds to 1 hour; ongoing
Thermodynamic changes from boiler control action (slow dynamics)	Seconds to hours
System structure monitoring (what is energized and what is not)	Steady state; on-going
System state measurement and estimation	Steady state; on-going
System security monitoring	Steady state; on-going
Load Management, load forecasting, generation scheduling	1 hour to 1 day or longer; ongoing
Maintenance scheduling	Months to 1 year; ongoing
Expansion planning	Years; ongoing
Power plant site selection, design, construction, environmental impact, etc.	2 years or longer

NERC; US Congress 2003). Other blackouts were experienced throughout the world in 2003 – in the UK, south London on August 28, in southern Sweden and parts of Denmark on September 23, and in Italy on September 28, affecting a total of approximately 112 million people. Earlier ones were in parts of France (December 1978) from a voltage collapse and destruction of high-voltage transmission towers due to high winds (December 1999), and massive outages in Canada, New York and New England from ice storms in January 1998 and December 1999. Other examples involve outages of infrastructure due to power failures. In April 1995, many regional trains suddenly stopped in Zürich, Switzerland when locomotives were automatically shut down by protective equipment because of excessively high harmonic currents from newly installed high-frequency converters. The number of such widespread cascading failures and their spill-over interdependencies into other critical infrastructures has been on the rise in recent years.

According to data from the North American Electric Reliability Council (NERC) and analyses from the Electric Power Research Institute (EPRI), average outages from 1984 to the present have affected nearly 700,000 North American customers annually. Smaller outages occur more frequently and affect tens to hundreds of thousands of customers every few weeks or months. Large outages occur every two to nine years and affect millions. Much larger outages affect seven million or more customers per decade – this rare event has considerably higher consequences. These analyses are based on data collected for the US Department of Energy (DOE), which requires electric utilities to report system emergencies including service interruptions, voltage reductions, sabotage, unusual occurrences that could affect the reliability of bulk power delivery systems, and fuel problems. (Amin 2003; Amin 2005; EPRI 1999–2003; EIA 2003; NERC).

Furthermore, today's electric power delivery system is based largely on technology developed in the 1950s or earlier, and installed over the last 30 to 50 years. The strain on this aging system is beginning to show. Figure 7.1 shows how major outages have affected electricity consumers. Generally, a relatively small number of US consumers experience a large number of outages; conversely, outages that affect a large number of consumers are rare. In Figure 7.1, the data point in the lower right-hand corner represents the widespread outage of August 10, 1996, which affected approximately seven million consumers in 11 western US states and two Canadian provinces. This plot could also indicate that the number of larger outages could be rising. Indeed, EPRI's analyses (EPRI 2003), found that 41 percent more outages affected 50,000 or more consumers in the second half of the 1990s than in the first half (58 versus 41). The average outage affected 15 percent more consumers from 1996–2000 than from 1991–1995 (409,854 versus 355,204). This can also be seen in terms of the large failures and load losses in Figure 7.2, as it shows that 76 outages led to a loss of 100 Mega Watts (MW) or more of power in the second half of the decade, compared to 66 such occurrences in the first half. During the same period, the average lost load caused by an outage increased by 34 percent, from an average lost load of 798 MW in the period from 1991 to 1995 to an average of 1067 MW from 1996 to 2000 (EIA 2003; NERC 1984–present; EPRI 2003).

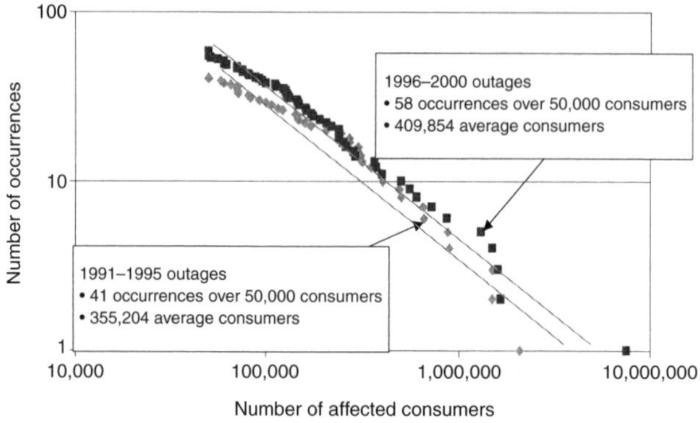

7.1 Number of US Power Outages Versus the Number of Consumers Affected, 1991 to 2000.

Source: Graphed from data courtesy of the North American Electric Reliability Council Disturbance Analysis Working Group database.

Note

Log-Log plot of outages and their impact on customers; our goal is to move these curves down toward the origin, i.e., to make outages less frequent and with smaller impact on customers.

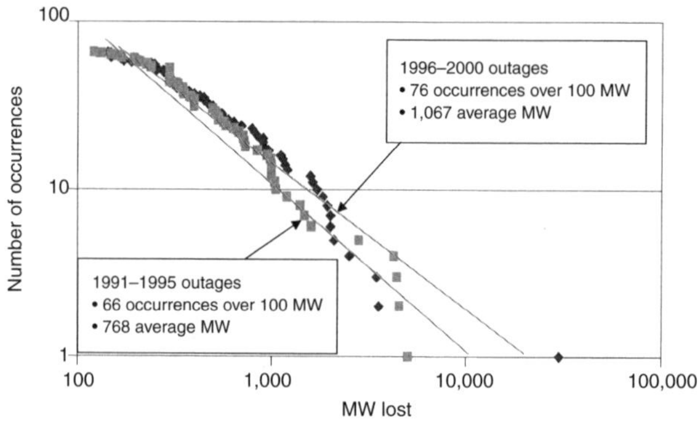

7.2 Number of US Power Outages Versus the Amount of Electric Load Lost, 1991 to 2000.

Source: Graphed from data courtesy of the North American Electric Reliability Council Disturbance Analysis Working Group database.

Note

Log-Log plot of outages and their size in mega watts.

Operational Systems Digital Network Control

IT has and will play a critical role in ensuring the reliable transmission and distribution of electricity. Electricity's share of total energy in the world is expected to continue to grow as more efficient and intelligent processes are introduced, such as controllers based on power electronics combined with wide-area sensing and management systems for improved performance. In the next two decades, it is envisioned that the electric power grid will move from an electro-mechanically controlled system to one that is electronically controlled.

In this sense, the electrical infrastructure is becoming increasingly intertwined with the IT infrastructure that supports it. Current and future power systems applications for telecommunications include the following:

- surveying overhead transmission circuits and rights-of-way
- transmitting SCADA system data (usually via telephone circuits)
- measuring overhead conductor sag
- measuring phasors (using a precise timing signal derived from the GPS to time-tag measurements of AC signals)
- fitting sine waves to AC signals, and determining magnitude and phase of v(t), i(t) in remote locations
- enhancing situational awareness by generating real-time pictures of system states and real-time power flow as well as real-time estimation of the systems' state and topology
- using data from Low Earth Orbit (LEO) satellites for faster-response control (more than 100 times less delay than High Earth Orbit (HEO) satellites) and connecting to existing parallel data stream facilities (effectively a high-speed global RS-232 channel).

The IT that supports the operational aspects and real-time control of electrical networks, range from energy management systems (EMS) to remote field devices. Critical systems include those described below.

Energy Management System (EMS). The objective of the EMS is to manage production, purchase, transmission, distribution and sale of electrical energy in the power system at a minimal cost with respect to safety and reliability. Management of the real-time operation of an electric power system is a complex task requiring interaction of human operators, computer systems, communications networks, and real-time data-gathering devices in power plants and substations. An EMS consists of computers, display devices, software, communication channels and remote terminal units that are connected to Remote Terminal Units (RTUs), control actuators, and transducers in power plants and substations. The main tasks it performs have to do with generator control and scheduling, network analysis, and operator training. Control of generation requires that the EMS maintain system frequency and tie line flows while economically dispatching each generating unit. Management of the transmission network requires that the EMS monitor up to thousands of telemetered values, estimate the electrical state of the network, and inform the operator of the best strategy to handle potential outages that could result in an overload or voltage limit violation. EMSs can have real-time two-way communication links between substations, power plants, independent system operators, and other utility EMSs.

Supervisory Control and Data Acquisition (SCADA) System. A SCADA system supports operator control of remote (or local) equipment, such as opening or closing a breaker. A SCADA system provides three critical functions in the operation of an electric power system: data acquisition, supervisory control, and alarm display and control. It consists of one or more computers with appropriate applications software connected by a communications system to a number of RTUs placed at various locations to collect data, perform intelligent control of electrical system devices, and report results back to an EMS. SCADAs can also be used for similar applications in natural gas pipeline transmission and distribution applications. A SCADA can have real time communication links with one or more EMSs and hundreds of substations.

Remote Terminal Unit (RTU). RTUs are special purpose microprocessor-based computers that contain analog to digital converters (ADC) and digital to analog converters (DAC), digital inputs for status and digital output for control. There are transmission substation RTUs and distribution automation (DA) RTUs. Transmission substation RTUs are deployed at substation and generation facilities where a large number of status and control points are required. DA RTUs are used to control air switches and VAR compensation capacitor banks (that support voltage) on utility poles, control pad-mounted switches, monitor and automate feeders, monitor and control underground networks and for various uses in smaller distribution substations. RTUs can be configured and interrogated using telecommunication technologies. They can have hundreds of real time communication links with other substations, EMS, and power plants.

Programmable Logic Controller (PLC). PLCs have been used extensively in manufacturing and process industries for many years and are now being used to implement relay and control systems in substations. PLCs have extended input/output (I/O) systems similar to transmission substation RTUs. The control outputs can be controlled by software residing in the PLC and via remote commands from a SCADA system. The PLC user can make changes in the software without making any major hardware or software changes. In some applications, PLCs with RTU reporting capability may have advantages over conventional RTUs. PLCs are also used in many power plant and refinery applications. They were originally designed for use in discrete applications like coal handling. They are now being used in continuous control applications such as feedwater control. PLCs can have many real-time communication links inside and outside substations or plants.

Protective Relays. Protective relays are designed to respond to system faults short circuits. When faults occur, the relays must signal the appropriate circuit breakers to trip and isolate the faulted equipment. Distribution system relaying must coordinate with fuses and reclosures for faults while ignoring cold-load pickup, capacitor bank switching and transformer energization. Transmission line relaying must locate and isolate a fault with sufficient speed to preserve stability, reduce fault damage, and minimize the impact on the power system. Certain types of "smart" protective relays can be configured and interrogated using telecommunication technologies.

Automated Metering. Automated metering is designed to upload residential and/or commercial gas and/or electric meter data. This data can then be automatically downloaded to a PC or other device and transmitted to a central collection point. With this technology, real-time communication links exist outside the utility infrastructure.

Plant Distributed Control Systems (DCSs). Plant Distributed Control Systems are plant-wide control systems that can be used for control and/or data acquisition. The I/O count can be as high as 20,000 data points or higher. Often, the DCS is used as the plant data highway for communication to/from intelligent field devices, other control systems such as PLCs, RTUs, and even the corporate data network for Enterprise Resource Planning (ERP) applications. The DCS traditionally has used a proprietary operating system. Newer versions are moving toward open systems such as Windows NT, Sun Solaris, etc. DCS technology has been developed with operating efficiency and user configurability as drivers, rather than system security. Additionally, technologies have been developed that allow remote access, usually via PC, to view and potentially reconfigure the operating parameters.

Field Devices. Examples of field devices are process instrumentation such as pressure and temperature sensor and chemical analyzers. Other standard types of field devices include electric actuators. Intelligent field devices include electronics to enable field configuration, upload of calibration data, etc. These devices can be configured off-line. They also can have real-time communication links between plant control systems, maintenance management systems, stand-alone PCs, and other devices inside and outside the facility.

Digital Interdependencies and Security Risks

Recognizing the increased interdependence between IT and electricity infrastructures, along with technical and business opportunities, electric power utilities typically own and operate at least parts of their own telecommunications systems which often consist of backbone fiber optic or microwave connecting major substations, with spurs to smaller sites. The energy industry has historically operated closed, tightly controlled networks. Deregulation and the resulting commercial influences have placed new information sharing demands on the energy industry. Traditional external entities like suppliers, consumers, regulators and even competitors now must have access to segments of the network. The definition of the network must be expanded to include the external wide area network connections for these external entities. This greatly increases the security risk to other functional segments of the internal network that must be protected from external connections. This is true whether a private network or the Internet is used to support the external wide area network.

The external entities already have connections to the Internet and as such the Internet can provide the backbone for the External Wide Area Network. Duplicating this backbone to create a private network requires not only large upfront start-up costs, but also ongoing maintenance costs and potentially higher individual transaction costs than using the Internet.

Information systems and on-line data processing tools include the Open Access Same-time Information System (OASIS), which is now in operation over the Internet; and Transfer Capability Evaluation (TRACE) software, which determines the total transfer capability for each transmission path posted on the OASIS network, while taking into account thermal, voltage, and interface limits.

Increased use of electronic automation raises issues regarding adequacy of operational security:

- reduced personnel at remote sites makes them more vulnerable to hostile threats
- interconnection of automation and control systems with public data networks makes them accessible to individuals and organizations, from any world-wide location using an inexpensive computer and a modem
- use of networked electronic systems for metering, scheduling, trading or e-commerce imposes numerous financial risks.

Utility telecommunications often include several media and diversified communications networks which in part provide redundancy; these range from dedicated fiber optic cables, digital and analog microwave, and VSAT satellite to power line carrier technology as well as the use of multiple address radio, spread spectrum radio, trunked mobile radio, and cellular digital packet data. Security of the cyber and communication networks now used by businesses is fundamental to the reliable operation of the grid; as power systems rely more heavily on computerized communications and control, system security has become increasingly dependent on protecting the integrity of the associated information systems. Part of the problem is that existing control systems, which were originally designed for use with proprietary, stand-alone communications networks, were later connected to the Internet (because of its productivity advantages and lower costs), but without adding the technology needed to make them secure. Communication of critical business information and controlled sharing of that information are essential parts of all business operations and processes.

As the deregulation of the energy industry unfolds, information security will become more important. Energy-related industries will have to balance what appears to be mutually exclusive goals of operating system flexibility with the need for security. Key electric energy operational systems depend on real-time communication links both internal and external to the enterprise. The functional diversity of these organizations has resulted in a need for these key systems to be designed with a focus on open systems that are user configurable to enable integration with other systems both internal and external to the enterprise. In many cases, these systems can be reconfigured for security using telecommunication technologies and in nearly all cases the systems dynamically exchange data in real time. Power plant DCS systems produce information necessary for dispatch and control. This requires real-time information flow between the power plant and the utility's control center, system dispatch center, regulatory authorities, etc. A power plant operating as part of a large wholesale power network may have links to an independent system operator, a power pool, etc. As the generation business moves more and more into market-driven competitive operation, both data integrity and confidentiality will become major concerns for the operating organizations.

Any telecommunication link which is even partially outside the control of the organization owning and operating power plants, SCADA systems or EMSs represents a potentially insecure pathway into business operations and to the grid itself. The interdependency analyses done by most companies during Y2K preparations have both identified these links and the systems vulnerability to their failures. Thus they provide an excellent reference point for a cyber-vulnerability analysis.

In particular, monitoring and control of the overall grid system is a major challenge. Existing communication and information system architectures lack coordination among various operational components, which usually is the cause for the unchecked development of problems and delayed system restoration. Like any complex dynamic infrastructure system, the electricity grid has many layers and is vulnerable to many different types of disturbances. While strong centralized control is essential to reliable operations, this requires multiple, high-data-rate, two-way communication links, a powerful central computing facility, and an elaborate operations control center, all of which are especially vulnerable when they are needed most – during serious system stresses or power disruptions. For deeper protection, intelligent distributed control is also required, which would enable parts of the network to remain operational and even automatically reconfigure in the event of local failures or threats of failure.

Distributed control capability is becoming available in next-generation integrated sensors that are equipped with two-way communication capability and support "intelligent agent" functions – not just sensing, but data assessment, adaptive learning, decision making, and actuation as well. The development of Intelligent Electronic Devices that combine sensors, telecommunication units, computers, and actuators will allow highly automated adjustments to be made at many points on the system and protect substantially against cascading failures. The use of distributed intelligent agents also opens the door to the development of a self-healing power grid that responds adaptively to counteract disturbances at the site of their occurrence.

Intelligent sensors will be capable of gathering a wide range of operating data, including time-stamped measurements of voltage, current, frequency, phase angle, and harmonics. This information, which provides input for distributed control, can also be integrated into a real-time system-wide database and coupled to analysis tools that perform dynamic monitoring, state estimation, disturbance analysis, and contingency assessment for the grid as a whole. Unfortunately, simulation-based techniques and mathematical models are presently unable to accurately portray the behavior of interactive networks whose dynamics can be highly nonlinear. Fine-tuning existing models with real-world input from distributed sensors may offer improvements, but substantial progress will require the formulation of new models.

SCADA and EMS system operations are critically dependent on the telecommunication links that gather data from geographically dispersed sources and transmit operational and control instructions to geographically dispersed facilities. In the North American grid these telecommunications links run the gamut from hardwired private networks to multi-network systems using a combination of private and public networks for both data acquisition and control. Not all of the networks are hardwired. Microwave and satellite communications links are common alternatives in areas where topography and/or distance makes wireless more cost effective. At first glance it would seem that a private, hardwired network which is totally within the control of the owner organization is a secure system. However even hardwired private networks will be linked to networks outside the control of the company. Typical outside data sources are bulk power customers, major retail customers, bulk power providers, power pools, independent system operating entities, etc. These connections can offer a multitude of paths into the SCADA and EMS systems. Without proper security design and management, each link is a potential security risk.

Challenges include how to handle network disruptions and delays. A major source of complexity is the interdependence of the telecommunication networks and the power grid. Issues range from the highest command and control level to the individual power stations and substations at the middle level, and then to the devices and power equipment at the lowest level.

Emerging Technologies

The introduction to this volume (Chapter 1) notes that technology is a two-edged sword. In the case of electricity, the aforementioned discussion reveals one edge (i.e. the risk) as the extent to which IT introduces a new set of security concerns. The other edge (i.e. the promise) remains the substantial increases in capacity and efficiency that are made possible through continuing IT advancements. The following is a sample of the emerging technologies that promise continuing gains in the electricity sector:

- Flexible Alternating Current Transmission System (FACTS) devices, which are high-voltage thyristor-based electronic controllers that increase the power capacity of transmission lines and have already been deployed in several high-value applications; at peak demand, up to 50 percent more power can be controlled than through existing lines

- Unified Power Flow Controller (UPFC), a third-generation FACTS device that uses solid-state electronics to direct power flow from one line to another to reduce overloads and improve reliability
- Fault Current Limiters (FCLs), which absorb the shock of short circuits for a few cycles to provide adequate time for a breaker to trip; preliminary results of analysis of the August 14, 2003, outage show that FCLs could have served as "shock absorbers" to limit the size of blackouts
- Innovations in materials science and processing, including high-temperature superconducting (HTS) cables, oxide-power-in-tube technology for HTS wire, and advanced silicon devices and wide-bandgap semiconductors for power electronics
- Information systems and on-line data processing tools such as the Open Access Same-time Information System (OASIS) and Transfer Capability Evaluation (TRACE) software, which determines total transfer capability for each transmission path posted on the OASIS network, while taking into account thermal, voltage, and interface limits
- Wide-Area Measurement Systems (WAMS), which integrate advanced sensors with satellite communication and time stamping using global positioning systems (GPS) to detect and report angle swings and other transmission system changes
- Enhanced IT systems for WAMS, OASIS, SCADA andEMS
- Advanced software systems for dynamic security assessment of large/wide-area networks augmented with market/risk assessment
- Intelligent Electronic Devices with security provisions built in by combining sensors, computers, telecommunication units, and actuators; related "intelligent agent" functions such as assessment, decision, and learning.

However, even if most of the above technologies are developed and deployed, there is still a major management challenge in making such a complex network perform reliably and with security. These issues are taken up next.

Management

Human Performance

The management chapter to this volume (Chapter 3) provides a framework that, among other items, illustrates that infrastructures are systems with "humans in the loop." This is indeed the case for electricity networks. Several key human resource issues arise in bringing IT to improvement the performance of electric power. The first is operator experience. The second is retaining professionals in the field of electric power engineering. The third is how users and consumers can interface with IT-enabled electric power systems.

Operator Training

Several root causes of the August 14 outage point to lack of operators' situational awareness and coordination. IT has a key role to play in the optimization of operator interfaces and other human factors issues. Basically, the problem is finding the most effective way for machines and humans to work together, and data glut and maintaining operator attention are largely at the center of the problem. Good operator interfaces provide adequate visualization of the state of the system, and they should be designed so the user can remain tuned in to many different factors while giving active attention to only a few.

Much of the answer is simply a matter of how information is packaged for viewing. IT innovations are expected to have applications in personnel training and optimization of human performance, for example through the use of virtual reality for training for maintenance or rapid repair work, especially involving hazardous situations. Voice recognition is another technology expected to come into broad use over the next decade; replacement of keyboarding with voice-based inputing could greatly streamline and simplify human interaction with computers and other electronic control equipment.

Since humans interact with these infrastructures as managers, operators and users, human performance plays an important role in their efficiency and security. In many complex networks, human participants themselves are both the most susceptible to failure and the most adaptable in the management of recovery. Modeling and simulating these networks, especially their economic and financial aspects, will require modeling the bounded rationality of "realistic" human thinking, unlike that of a hypothetical "expert" human as in most applications of artificial intelligence (AI). Even more directly, most of these networks require some human intervention for their routine control, especially when they are exhibiting anomalous behavior that may suggest actual or incipient failure.

Retaining a Trained Workforce

A growing concern related to the human network is the erosion of technical knowledge within the power industry. To a large extent this is a matter of the retirement of seasoned power engineers, exacerbated by recent downsizing reductions of in-house workforces. These key employees take their knowledge with them when they go. It will take a long time to recruit replacements. A second related issue is that new engineers are not entering the field rapidly enough to replace retirees. The average power engineer's age has increased significantly over the last two decades. A serious shortage of power engineers is developing, and is expected to continue for several decades.

Users

Operators and maintenance personnel are obviously "inside" these networks and can have direct, real-time effects on them. But users of a telecommunication, transportation, electric power or pipeline system also affect the behavior of those systems, often without conscious intent. The amounts, and often nature, of demands put on the network can be the immediate cause of conflict, diminished performance and even collapse. Reflected harmonics from one user's machinery degrade power quality for all. Long transmissions from a few users create Internet congestion. Simultaneous lawn watering drops everyone's water pressure. No one is "outside" the infrastructure.

Given that there is some automatic way to detect actual or imminent local failures, the obvious next step is to warn the operators. Unfortunately, the operators are usually busy with other tasks, sometimes even responding to previous warnings. In the worst case, detected failure sets off a multitude of almost simultaneous alarms as it begins to cascade through the system, and before the operators can determine the real source of the problem, the whole network has shut itself down automatically.

Unfortunately, humans have cognitive limitations that can cause them to make serious mistakes when they are interrupted. In recent years, a number of systems have been designed that allow users to delegate tasks to intelligent software assistants ("softbots") that operate in the background, handling routine tasks and informing the operators in accordance with some protocol that establishes the level of their delegated authority to act independently. In this arrangement, the operator becomes a supervisor, who must either cede almost all authority to subordinates or be subject to interruption by them. At present, we have very limited understanding of how to design user interfaces to accommodate interruption.

Information Security

The electric power industry traditionally has been a vertically integrated industry that in some cases operated in pseudo-monopolistic fashion. However, the industry is currently undergoing restructuring, frequently resulting in a break-up of the vertical structure. Additionally, there has been a significant move on the part of control system suppliers to electric and petrochemical industries toward open, user-configurable systems utilizing real-time communications. With a vertical structure, local and wide-area networks were sufficient to maintain a reasonably secure data network. However, deregulation and new networking technologies are making secure communications more important, and more difficult to develop and maintain.

Information security is concerned with the relationships between people and information. In these relationships, people are owners, custodians, creators, readers, modifiers, certifiers, or even subjects of the information. It follows then that the information itself is the object of various actions by people – creation, destruction, reading, modification, and certification. Information security is concerned with first defining appropriate relationships between people as actors and information resources as objects; these relationships are usually defined as a set of rules defining permitted actions. Not all threats come from outside the organization nor are all threats malicious.

Information security is also concerned with controlling the relationships between people and information so that information is managed according to well-defined rules. Some human agent or institutional agency of authority is usually charged with creating, communicating, applying, monitoring and enforcing these information security protocols. Examples of contemporary information security rules are: rules for handling government classified documents; rules for ensuring client-attorney privilege or privacy of shared information; rules followed by corporate accountants and checked by financial auditors; and rules for ensuring accuracy and completeness of patients' health records. Generally these rules define information security controls based on properties of special classes of information; these properties fall into three broad categories: confidentiality of sensitive information; integrity and authenticity of critical information; and availability of necessary information. These principles need to be applied to the management of electricity systems, including the operators and managers of these systems.

Complex System Failure

Beyond the human dimension, there is a strategic need to understand the societal consequences of infrastructure failure risks along with benefits of various tiers of increased reliability. From an infrastructure interdependency perspective, power, telecommunications, banking and finance, transportation and distribution, and other infrastructures are becoming more and more congested, and are increasingly vulnerable to failures cascading through and between them. A key concern is the avoidance of widespread network failure due to cascading and interactive effects. Moreover, interdependence is only one of several characteristics that challenge the control and reliable operation of these networks. Other factors that place increased stress on the power grid include dependencies on adjacent power grids (increasing because of deregulation), telecommunications, markets, and computer networks. Furthermore, reliable electric service is critically dependent on the whole grid's ability to respond to changed conditions instantaneously.

Prior to the tragic events of September 11, the US President's Commission on Critical Infrastructure Protection in October, 1997, highlighted growing concern (CIAO 1997). It noted the damaging and dangerous ways cascading failures could unpredictably affect the economy, security, and health of citizens. Secure and reliable operation of these systems is fundamental to our economy, security and quality of life, as noted in the report and the subsequent Presidential Directive 63 on Critical Infrastructure Protection, issued on May 22, 1998.

Secure and reliable operation of critical infrastructures poses significant theoretical and practical challenges in analysis, modeling, simulation, prediction, control, and optimization. To address these challenges, a research initiative – the EPRI/DOD Complex Interactive Networks/Systems Initiative (CIN/SI) – was undertaken during 1998–2001 to enable critical infrastructures to adapt to a broad array of potential disturbances, including terrorist attacks, natural disasters, and equipment failures.

The CIN/SI overcame the longstanding problems of complexity, analysis, and management for large interconnected systems – and systems of systems – by opening up new concepts and techniques for the strategic management of this infrastructure system. Dynamical systems, statistical physics, information and communication science, and computational complexity were extended to provide practical tools to measure and model the power grid, cell phone networks, Internet, and other complex systems. For the first time, global dynamics for such systems could be understood fundamentally.

One example of this work can be seen in Figure 7.3, which charts the economic losses from disasters. Economic losses from disasters were found to follow a power law distribution – for hurricanes, floods, earthquakes, and even electrical outages. In other words, higher loss events tended to occur less frequently. Fundamental power law distributions also were found for forest fires, Internet congestion, and other systems. CIN/SI results such as these translate in new approaches

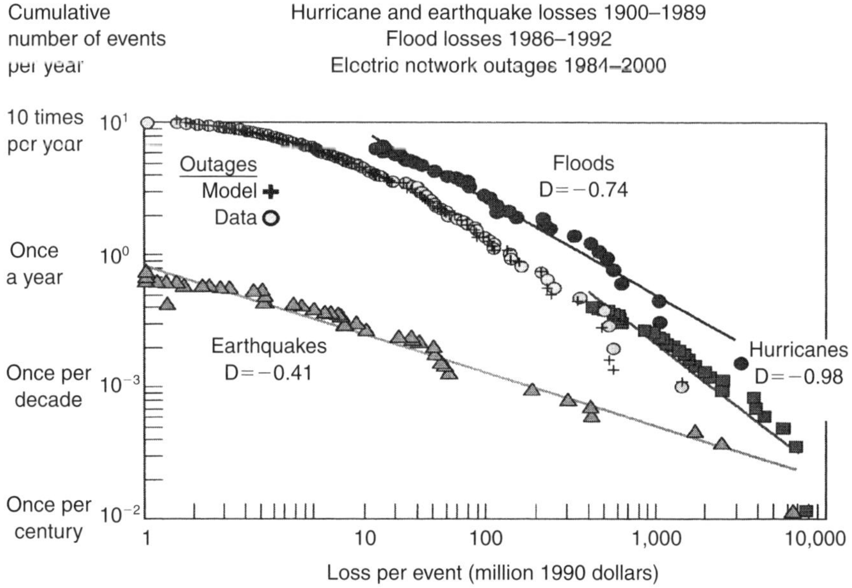

7.3 *Understanding Complex Systems and Global Dynamics.*
Source: The EPRI/DoD Complex Interactive Networks/Systems Initiative (CIN/SI).

for optimizing complex systems in terms of productivity and robustness to disaster. It is the intent of this research program to use these findings to aid in devising disaster prevention by developing response and resolution strategies that managers can use to minimize the economic consequences of such disasters.

CIN/SI has also developed a strategic vision for the integrated sensing, communications, and control of the power grid. Some pertinent management issues are whether to emphasize centralized or decentralized control and how to ensure adaptive operation and robustness to disturbances. These include various types of failures ranging from intentional to inadvertent (Amin 2000a; Amin 2000b; Amin 2001; EPRI 2004). In short, this roadmap aims to refine not only how IT can enhance the power grid, but the management challenges of doing so.

Policy

Impact of Deregulation

The most pervasive policy influence on the electrical infrastructure in the United States (as well as in several other industrialized countries) in recent years has been the ongoing deregulation of systems and services. While more populous countries like China and India will have greater potential electricity markets and demands (see Chapter 10), the United States presently comprises the largest national market for electric power. Its electric utilities have been mostly privately owned, vertically integrated and locally regulated. National regulations in areas of safety, pollution and network reliability also constrain their operations to a degree, but local regulatory bodies, mostly at the State level, have set prices and return on investment, and have controlled investment decisions while protecting utilities from outside competition. That is now rapidly changing. State regulators are moving toward encouraging a competitive electric power market.

The electric power grid was historically operated by separate utilities, each independent in its own control area and regulated by local bodies, to deliver bulk power to load areas reliably and economically. As a non-competitive, regulated monopoly, emphasis was on reliability (and security) and economy. Competition and deregulation have created multiple energy producers that must share the same regulated energy delivery network. Traditionally, new delivery capacity would be added to handle load increases, but because of current difficulty in obtaining permits and uncertainty about achieving an adequate rate of return on investment, circuit miles added annually are declining while demand for delivery resources grows. In recent years, the "shock absorbers" have been shrinking. The network is becoming increasingly stressed, and whether the carrying capacity or safety margin will exist to support anticipated demand is in question. The complex systems used to relieve bottlenecks and clear disturbances during peak demand are at great risk of serious disruption, creating a critical need for technological improvements.

It is important to note that while network failure can occur instantaneously, a very different time horizon is needed to understand the underlying problems and plan for their resolution. This interval spans years of inadequate investments that have created a fundamental imbalance in the infrastructure, its reliability and markets/policies impacting it. The imbalance is prevalent throughout the nation. In addition, these and other national and international policy and economic changes are impacting energy security and compounding uncertainties. In terms of resolution, public policy and private markets need to encourage long term investment in the electricity infrastructure so that electricity supply and demand can be instantaneously balanced at all times. Indeed, it is a paradox that a real-time infrastructure such as electricity requires a long-term policy view even in a deregulated policy environment.

Infrastructures Under Threat

The terrorist attacks of September 11 exposed critical vulnerabilities in America's essential infrastructures: the security of these fundamental systems can no longer be taken for granted. Electric power systems constitute *the* fundamental infrastructure of modern society. A successful terrorist attempt to disrupt electricity supplies could have devastating effects on national security, the economy, and the lives of every citizen. Yet power systems have widely dispersed assets that can never be absolutely defended against a determined attack.

The growing potential for infrastructure problems stems from multiple sources, including system complexity, deregulation, economic effects, power-market impacts, terrorism, and human error. The existing power system is also vulnerable to natural disasters.

A November 2001 EPRI assessment developed in response to the September 11 attacks highlights three kinds of potential threats to the US electricity infrastructure (Amin 2000–2003; EPRI 2001). We discuss them briefly and in very broad brushes, without providing a "blue book" for potential attackers: the first is attacks *upon* the power system. In this case, electricity infrastructure is the primary target – with ripple effects, in terms of outages, extending into the customer base. The point of attack could be a single component, such as a critical substation or a transmission tower. There could also be a simultaneous, multi-pronged attack intended to bring down the entire grid in a region of the US An attack could also target electricity markets, which are highly vulnerable because of their transitional status.

The second type of attack is *by* the power system. In this case, the ultimate target is the population, using parts of the electricity infrastructure as a weapon similar to the way our transportation and mail delivery systems were used against our nation. Power plant cooling towers, for example, could be used to disperse chemical or biological agents.

The third means is attack *through* the power system. In this case, the target is the civil infrastructure. Utility networks include multiple conduits for attack, including lines, pipes, underground cables, tunnels and sewers. An electromagnetic pulse, for example, could be coupled through the grid with the intention of damaging computer and/or telecommunications infrastructure.

As seen from these scenarios, the specter of terrorism raises a profound dilemma for the electric power industry: how to make the electricity infrastructure more secure without compromising productivity advantages inherent in today's complex, highly interconnected electric networks? Resolving this dilemma requires short-term and long-term technology development and deployment, affecting fundamental characteristics of today's power systems.

The North American electric power system needs a comprehensive strategy to prepare for the diverse threats posed by terrorism. Such a strategy should both increase protection of vital industry assets and assure the public that they are well protected. A number of actions will need to be considered in formulating an overall security strategy:

- the grid must be made secure from cascading damage
- pathways for environmental attack must be sealed off
- conduits for attack must be monitored, sealed off and "sectionalized" under attack conditions
- critical controls and communications must be made secure from penetration by hackers and terrorists
- greater intelligence must be built into the grid to provide flexibility and adaptability under attack conditions, including automatic reconfiguration
- ongoing security assessments, including use of game theory to develop potential attack scenarios, will be needed to ensure that the power industry can stay ahead of changing vulnerabilities.

A survey of electric utilities revealed real concerns about grid and communications security. Figure 7.4 ranks perceived threats to utility control centers. The most likely threats were bypassing controls, integrity violations, and authorization violations, with four-in-ten rating each as either a five, or four out of five. Concern about potential threats generally increased as the size of the utility (peak load) increased.

The system's equipment and facilities are dispersed throughout the North American continent, which complicates absolute protection of the system from a determined terrorist attack. In addition, another complexity needs to be considered – the power delivery systems' physical vulnerabilities and susceptibility to disruptions in computer networks and communication systems. For example, terrorists might exploit the increasingly centralized control of the power delivery system to magnify effects of a localized attack. Because many consumers have become more dependent on electronic systems that are sensitive to power disturbances, an attack that leads to even a momentary interruption of power can be costly. A 20-minute outage at an integrated circuit fabrication plant, for example, could cost US$30 million.

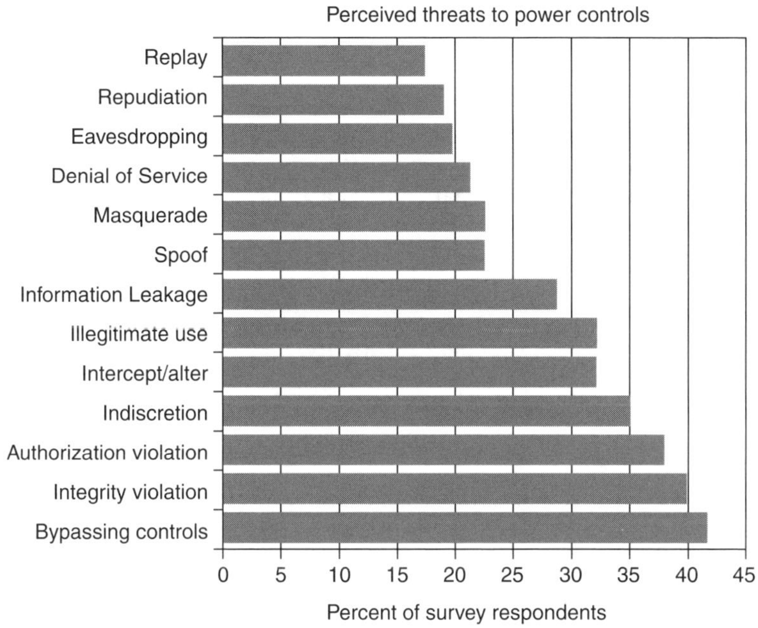

7.4 *Threats to Power Supply Control.*
Source: Graphed from EPRI (2000b).

Conclusion

A Stressed Infrastructure

The major outage on August 14, 2003 in the Eastern US and the earlier California power crisis in 2000–2001 are only the most visible parts of a larger and growing US energy crisis stemming from inadequate investments in the infrastructure leading to a fundamental imbalance between growing demand and an almost stagnant supply. The imbalance existed for many years and is prevalent throughout the nation (EPRI 2001).

From a broader view, the North American electricity infrastructure is vulnerable to increasing stresses from several sources. One stress is caused by an imbalance between growth in demand for power and enhancement of the power delivery system to support this growth. From 1988–1998, the US's electricity demand rose by nearly 30 percent, but the capacity of its transmission network grew by only 15 percent. This disparity will likely increase from 1999–2009: analysts expect demand to grow by 20 percent, while planned transmission systems grow by only 3.5 percent. Along with that imbalance, today's power system has several sources of stress:

- *Demand is outpacing infrastructure expansion and maintenance investments.* Generation and transmission capacity margins are shrinking and unable to meet peak conditions, particularly when multiple failures occur while electricity demand continues to grow.

- *The transition to deregulation is creating new demands that are not being met.* The electricity infrastructure is not being expanded or enhanced to meet the demands of wholesale competition in the industry so connectivity between consumers and markets is at a gridlock.

- *The present power delivery infrastructure cannot adequately handle the new demands of high-end digital customers and the twenty-first century economy.* It cannot support levels of security, quality, reliability, and availability needed for economic prosperity.

- *The infrastructure has not kept up with new technology.* Many distribution systems have not been updated with current technology including IT.

- *Proliferation of distributed energy resources (DER).* DER includes a variety of energy sources – micro turbines, fuel cells, photovoltaics, and energy storage devices – with capacities from approximately 1 kW to 10 MW. DER can play an important role in strengthening energy infrastructure. Currently, DER accounts for about seven percent of total capacity in the United States, mostly in the form of backup generation, yet very little is connected to the power delivery system. By 2020, DER could account for as much as 25 percent of total US capacity, with most DER devices connected to the power delivery system.

- *Return on investment (ROI) uncertainties are discouraging investments in the infrastructure upgrades.* Investing new technology in the infrastructure can meet the aforementioned demands. More specifically, according to a June 2003 report by the National Science Foundation, R&D spending in the US as a percent of net sales was about ten percent in the computer and electronic products industry and 12 percent for the communication equipment industry in 1999. Conversely, R&D investment by electric utilities was less than 0.5 percent during the same period. R&D investment in most other industries is significantly greater than that in the electric power industry (NSF 2003).

- *Concern about the electrical infrastructure's security* (Amin 2003; EPRI 2001). A successful terrorist attempt to disrupt electricity supplies could have devastating effects on national security, the economy, and human life. Yet power systems have widely dispersed assets that can never be absolutely defended against a determined attack.

Competition and deregulation have created multiple energy producers that share the same energy distribution network, one that now lacks the carrying capacity or safety margin to support anticipated demand. Investments in maintenance and research and development continue to decline as the North American electrical grid. Yet, investment in core systems and related IT components are required to unsure the level of reliability and security that users of the system have come to expect.

Next Steps

Digital Technology can make a vital contribution to reliability and security by enhancing power systems' inherent resilience and flexibility to withstand terrorist attacks and natural disasters. Creating a smart grid with self-healing capabilities is no longer a distant dream; we've made considerable progress. But considerable technical challenges and several economic and policy issues remain to be addressed, including:

- What threat level is the industry responsible for?
- What does government need to address?
- Will market-based priorities support a strategically secure power system?
- Who will pay for it and what are the economic incentives for such investments?
- What overall system architecture is most conducive to maintaining security?

Energy policy and technology development require long-term commitments as well as sustained and patient investments in technology creation and development of human capital.

To address these and other vulnerabilities, the electric power industry and all pertinent public and private sectors must work together with other critical infrastructure stakeholders. Specifically, we should consider carrying out a recommendation in the National Research Council (2002) report, "Making the Nation Safer: The Role of Science and Technology in Countering Terrorism":

> A coordinating council should be formed to ensure that the necessary research on electric power systems is carried out, that the resulting technologies have a route to market, that implementation is done expeditiously, and that the costs are recovered through appropriate incentives, fees, rate adjustments, or other funding mechanisms. The council should include, but not be limited to, representation from the North American Electric Reliability Council, DOE, the Office of Homeland Security, NARUC, EPRI and other utility industry groups, manufacturers, and ISOs and RTOs.

As additional technology or policy resources, there are pertinent strategic plans and roadmapping activities for electric delivery technologies, including: US DOE's National Transmission Grid Study (DOE 2002) and the on-going roadmap activity at EPRI (EPRI 1999; EPRI 2003), as well as in NRECA and the California Energy Commission. Just one week prior to the blackout, EPRI released a report on the challenges facing the electricity sector in the US, outlining a Framework for Action, entitled, the Electricity Sector Framework for the Future (ESFF).

Toward High Performance Digital Systems

Despite increasing concerns in these areas as well as the continuing erosion of reserve margins, today's infrastructures generally function well. The electricity is on, the phones work, traffic flows nearly all of the time. But more and more, the traditional level of performance is no longer good enough; more robust infrastructures are needed for the "Digital Society" envisioned for tomorrow. In the electric power area, there is a need for an increase in reliability from today's average of about 99.9 percent (approximately 8 hours of outage per year) to 99.9999 percent (about 32 seconds outage per year) or even 99.9999999 percent (one outage lasting less than a single AC cycle per year) to reach the level of performance that various customers may require. IT holds the promise to achieve such reliability, and also directly benefits from these improvements. Such near-perfect power is needed for error-free operation of the microprocessor chips finding their way into just about everything, including billions of embedded applications, these days.

Tomorrow's Digital Society will be built on microprocessors – in smart refrigerators that automatically keep themselves stocked with your favorite foods, in smart door locks that know who to let in and when, smart shoes that monitor your daily exercise and physical condition, smart wallets, smart curtains, smart toothbrushes, smart cereal boxes. Adequate microprocessors now cost just a few dollars; before long the price will be a few cents or less. The catch is the quality of electricity service: unprotected microprocessors demand perfect power to function properly. A similar need for perfection exists for other infrastructures, where future advanced systems are predicated on the perfect functioning of today's communications, transportation, and financial services. Given economic, societal, and quality-of-life issues and the ever-increasing interdependencies among infrastructures, a key challenge before us is whether the electricity infrastructure will evolve to become the primary support for the twenty-first century's digital society – a smart grid with self-healing capabilities – or be left behind as a twentieth century industrial relic?

Acknowledgments

I thank editors of this volume, Professors Rae Zimmerman and Tom Horan, for their encouragement and continued interest in this subject. I developed most of the material and findings presented here while I was at the Electric Power Research Institute (EPRI) in Palo Alto, California. I gratefully acknowledge EPRI's support, and feedback from numerous colleagues at EPRI, universities, industry, and government agencies.

References

Amin, M. (2000a) "Toward Self-Healing Infrastructure Systems," *IEEE Computer Magazine*, Vol. 33, No. 8, August, pp. 44–53.

Amin, M. (2000b) "National Infrastructures as Complex Interactive Networks," Chapter 14 in Samad and Weyrauch (eds) *Automation, Control, and Complexity: New Developments and Directions*, John Wiley and Sons, March.

Amin, M. (2001) "Toward Self-Healing Energy Infrastructure Systems," cover feature in the *IEEE Computer Applications in Power*, Vol. 14, No. 1, January, pp. 20–8.

Amin, M. (2002a) "Security Challenges for the Electricity Infrastructure," special issue of the *IEEE Computer Magazine on Security and Privacy*, April.

Amin, M. (2002b) Special issues of *IEEE Control Systems Magazine on Control of Complex Networks*, Vol. 21, No. 6, December 2001, and Vol. 22, No. 1, February.

Amin, M. (2003) "North America's Electricity Infrastructure: Are We Ready for More Perfect Storms," *IEEE Security and Privacy Magazine*, Vol. 1, No. 5, September/October, pp. 19–25.

Amin, M. (2005) Special Issue of the Proceedings of the IEEE on Energy Infrastructure Defense System, forthcoming, November.

CIAO (1997) "Critical Foundations: Protecting America's Infrastructures, The Report of the President's Commission on Critical Infrastructure Protection," US Department of Commerce Critical Infrastructure Assurance Office (CIAO), October, Washington, D.C. Online. Available at. www.ciao.gov/resource/pccip/PCCIP_Report.pdf (accessed January 14, 2004).

DOE (2002) National Transmission Grid Study, US Department of Energy (DOE), May, Online. Available at: tis.eh.doe.gov/ntgs/gridstudy/main_screen.pdf (accessed January 14, 2004).

Dy Liacco, T.E. (1967) "The Adaptive Reliability Control System," *IEEE on PAS*, May, pp. 517–61.

EIA (2003) *Annual Energy Outlook*, US Department of Energy, Energy Information Administration (EIA), Online: Available at: www.eia.doe.gov/oiaf/aeo/figure_3.html (accessed January 14, 2004).

EPRI (1999) "Electricity Technology Roadmap: 1999 Summary and Synthesis," Technical Report, CI-112677-V1, 160 pp., Palo Alto, CA, July, Online. Available at: www.epri.com/corporate/discover_epri/roadmap/index.html (accessed January 14, 2004).

EPRI (2000a) "Complex Interactive Networks/Systems Initiative: Overview and Progress Report for Joint EPRI/Dept. of Defense University Research Initiative," First year progress report TP-114660, Palo Alto, CA, May.

EPRI (2000b) "Communication Security Assessment for the United States Electric Utility Infrastructure," Report 1001174, December, pp. 4–11.

EPRI (2001) "Electricity Infrastructure Security Assessment," Vol. I-II, Palo Alto, CA, November and December.

EPRI (2003) "Electricity Technology Roadmap: Synthesis Module on Power Delivery System and Electricity Markets of the Future," Palo Alto, CA, July.

EPRI (2004) "Complex Interactive Networks/Systems Initiative: Overview and Summary," Final

Report for Joint EPRI and US Department of Defense University Research Initiative, Palo Alto, CA, March.
Fink, L.H. and Carlsen, K. (1978) "Operating Under Stress and Strain," *IEEE Spectrum*, March, pp. 48–53.
Kundur, P. (1994) *Power System Stability and Control*, EPRI Power System Engineering Series, McGraw-Hill, Inc.
National Research Council (NRC) (2002) *Making the Nation Safer: The Role of Science and Technology in Countering Terrorism*. Washington, D.C.: National Academy Press.
National Science Foundation (NSF) (2003) "Research and Development in Industry: 2000," National Science Foundation, Division of Science Resources Statistics (DSRS), Arlington, VA (NSF 03-318), June, Online. Available at: www.nsf.gov/sbe/srs/nsf03318/pdf/taba19.pdf (accessed January 14, 2004).
NERC (1984–present) North American Electric Reliability Council (NERC) Disturbance Analysis Working Group (DAWG) database, Online. Available at: www.nerc.com/~dawg/ (accessed January 14, 2004).
North American Electric Reliability Council (NERC) at http://www.nerc.com/
US Congress (2003) Committee hearing of the House Committee on Energy and Commerce, "Blackout 2003: How Did It Happen and Why?" September 3–4, Online. Available at: energycommerce.house.gov (accessed January 14, 2004).

CHAPTER EIGHT

Telecommunications: Catastrophe and Recovery in the Information City

Mitchell Moss and Anthony Townsend

Introduction

Telecommunications is central to modern life and to the performance of other infrastructure that supports these lifestyles. The events of September 11, 2001 reflected not only how telecommunications systems operate as an infrastructure in a time of crisis, but also the magnitude of their support of basic economic functions and services, including other infrastructure.

Since the invention of the telegraph over 150 years ago, lower Manhattan has been the world's leading telecommunications hub. Telecommunications technologies evolved symbiotically with the financial services industry, providing the means to quickly and cheaply communicate with colleagues across the city or across the globe. As a result, there is more fiber optic cable strung beneath the streets of Manhattan than on the entire continent of Africa. The two main telephone switches in the Financial District each house more lines than many European nations. The ether above is saturated with an endless variety of wireless transmissions, such as those that emanated from the more than 1,500 antenna structures mounted atop the World Trade Center's north tower.

The September 11, 2001 terrorist attacks sparked the largest telecommunications event in human history and in the process focused attention on the pervasiveness of digital network infrastructure in today's cities. While transportation, water, and power networks are all critical to the proper functioning of a modern metropolis, during crisis and times of uncertainty communications networks play a crucial role in urban survival.

This chapter examines the events of September 11, 2001 and the following months in New York City by focusing on the role of digital communications networks. In particular, it describes three phases of the disaster in which digital networks played a major role – initial response, interim restoration, and long-term recovery. During each of these periods, it shows how digital network infrastructure, while the most fragile of all urban networks, is remarkably resilient in times of crisis.

Initial Response

Almost as soon as the first plane struck 2 WTC at 8:48 am, telecommunications systems throughout the world began experiencing what would soon become the biggest surge of use in history. As many have pointed out, New York's role as a media center was undoubtedly one reason it was chosen as a target – it was almost certain that video cameras would inadvertently capture the attacks live (Graham 2001).

Even before emergency workers had begun to respond to the raging high-rise fire caused by the impact of the first plane, television networks and news websites were pumping images of urban catastrophe across the globe. While most people choose the richness and immediacy of broadcast media such as radio and television over the Internet, popular news websites were swamped with request for images and streaming video.[1] Traffic to the CNN website doubled every seven minutes between 9 and 10 am that morning until its servers were no longer able to keep up with demand and crashed. Even after stripping their site down to the so-called "end of the world page," CNN could not keep up with an estimated 2 million page requests per minute (NRC 2003).

The telephone network was under severe strain. Tens of millions of Americans instinctively reached for their telephones as news of the plane crashes began to spread, and historical call volumes were sustained throughout much of the week. On September 11 alone, AT&T connected some 431 million calls, 20 percent more than normal. Based on AT&T's long-distance market share, it is likely that over 700 million long-distance calls were connected nationwide that day. Approximately double this number of calls were attempted, but AT&T's experience with previous localized disasters such as earthquakes and hurricanes had prepared them. Giving priority to the needs of the affected people, AT&T blocked incoming calls to keep outgoing lines from New York and Washington open. Local telephone systems were swamped as well, with Verizon reporting over 250 million local phone calls in the New York region in the week of September 11–18. Phone traffic in New York City was at double the normal levels, an unprecedented level (NRC 2003).

Response Within New York City

Government Response

While millions watched helpless and tried to reach loved ones, the men and women of the NYPD and NYFD, along with state and federal authorities, were rushing into the burning buildings, evacuating lower Manhattan for the second time in a decade, and cordoning off what had quickly been transformed from disaster area to crime scene after the impact of the second plane at 9:03 am. The failure of mobile wireless emergency communications that these brave men and women experienced is perhaps the most tragic part of this story. The breakdowns fell into two main categories. First, the repeater systems for fire and police radios (installed inside the World Trade Center towers after the 1993 terrorist bombing) failed when power was cut after the planes' impact. Few can forget the voices of firefighters captured on documentary footage as they desperately try to understand the weak voices coming through their radios. A report prepared by McKinsey found that problems with the radio system put many of the fire companies out of touch as they ascended the stairs into the twin towers. As a result, at least 120 firefighters did not receive the order to evacuate and were killed in the second collapse (Dwyer *et al.* 2002).

Second, even when radio equipment worked perfectly, first responders were hindered due to the lack of inter-operable radios. Supporting firefighters arriving from New Jersey were unable to communicate by radio with their New York City counterparts, who used a different communications frequency. Other problems, caused by psychological factors rather than technological ones, also strained the capacity of emergency radio channels. The system used by New York City's Emergency Medical Service was swamped with unnecessary transmissions from panicked operators, causing congestion that interfered with important messages.

The sheer variety of first responders further contributed to the breakdown of communications. Dozens of local, state, and federal authorities responded to the attacks. Yet, since few law enforcement agencies share compatible communications technologies, many rely upon public networks like the cellular telephone system for inter-agency communications. Throughout the day, congestion plagued the public cellular systems as historical call volumes combined with oversubscribed networks. During the morning of September 11, fewer than 1 in 20 mobile calls were connected in New York City. Since no system was in place to prioritize calls for emergency workers, response was confused and uncoordinated throughout much of the day.

The Financial Industry Response

While law enforcement and local government responded haphazardly due to a lack of effective communications, the financial services industry responded rapidly and decisively to the attacks and resulting destruction. This quick and capable response was the result of preparations dating back to the 1993 World Trade Center bombing and beyond.

One of the most severely affected firms, Lehman Brothers, responded almost instantly. Lehman CTO Bob Schwartz was able to activate the company's disaster recovery plan just after 9 am by sending a message from his portable Blackberry messenger while descending the stairwell of 1 WTC. While Lehman's Manhattan data center was destroyed in the collapse of the towers, operations were seamlessly transferred to a twin backup center in New Jersey. The firm was ready for trading at backup facilities in New Jersey the next day, five days before the stock markets were ready to reopen (Gaudin 2001).

While Lehman Brothers was one of the most severely affected firms, it was by no means unique. By noon on the 11th, every major bank in New York City had activated some form of disaster recovery plan. By the next morning, nearly 50 firms in and around the World Trade Center had declared disasters and called upon their disaster recovery contractors – firms such as SunGard and Comdisco – to retrieve sensitive data from off-site backups. Many firms (like Lehman Brothers) were able to switch over instantaneously to backup systems in other locations – mainly in midtown Manhattan, New Jersey, and downtown Brooklyn (Ohlson 2001). Cantor-Fitzgerald, which lost some 700 employees in the WTC, was up and running when the bond market reopened on Thursday morning, 47 hours later. This was possible because the firm had duplicate data centers in both New Jersey and London (Cone *et al.* 2001).

National Trends

Outside New York City, September 11 was marked by abrupt shifts in telecommunications use. As the afternoon wore on, telephone traffic began to subside and many turned to the Internet for more detailed information. Around 4pm, a surge of users rushed to AOL and other online services, flooding the Internet with short mails and instant messages. Presumably, once closest contacts had been reached by telephone, people were broadcasting their whereabouts and status through blanket mailings to their entire address book. AOL reported an astonishing 1.2 billion instant messages per day throughout September, far higher than before the 11th.

Across the Internet, there were similar patterns. The AP photo archive on AOL, which had only received 30,000 requests daily before the attacks, received over one million daily afterwards. Content distribution networks like Akamai observed a three- to four-fold increase in streaming media demand. Ironically, however, the insatiable demand for Internet pornography could not be deterred – "World Trade Center" still ranked just seventh among keyword searches on AOL, behind #1 "Nostradamus" and five pornographic terms!

As quickly as Internet use had surged, it faded away as President Bush took the stage late in the evening to calm a frightened nation. Much like during major television events such as the Super Bowl, AOL and other ISPs recorded record low usage during the President's speech. Millions logged off the Internet and gathered around television sets to hear how their leaders would react.

Interim Restoration

Restoration occurred across the vast telecommunications system that had evolved over a century. The restoration of the telephone system, the Internet, and wireless systems illustrates the extensiveness of the restoration effort in the days and weeks following the attacks. The restoration of the Stock Exchange illustrates how multiple telecommunications networks were brought back to revitalize a key global financial institution.

The Telephone System

The second phase of telecommunications activity began in earnest on September 12 as the extent of damage to the city's digital network infrastructure became clear. Direct physical damage to the city's telecommunications infrastructure was extensive. The facilities destroyed included:

- 3.5 million local phone lines were served by Verizon from their 140 West Street central office (directly across from the WTC), which was severely damaged by falling debris, smoke, and water.

- AT&T lost a large central office it operated in the basement of the WTC complex. Remarkably however, the equipment was undamaged during the building collapse and continued to operate until it lost battery power at 4 pm on September 11. This switch served some 20,000 T1 lines and over 1,200 T3 lines in the WTC area.

- 1,500 antennas atop 2 WTC including most major TV and radio broadcasters were lost.

- At least 15 *cellular telephone antenna sites* were lost, severely constraining local wireless capacity in and around Ground Zero. Many other cell sites were left intact, but their landline connections into the grid were knocked out by the damage at 140 West Street.

Rebuilding the local phone system would cost Verizon some US$1.4 billion, take more than 12 months, and involve some 3,000 technicians and engineers. AT&T spent an undisclosed amount in the "hundreds of millions" to replace facilities it lost in the attacks.

Meanwhile, the surviving digital networks would be called upon to help revive the city's morale and ensure the continuing operation of the financial markets – the lifeblood of the city and a crucial lynchpin of the national economy. While digital networks were proved to be very fragile on September 11, the sheer diversity of overlapping systems and the self-healing properties increasingly integrated have shown them to be remarkably resilient. This resiliency, in turn, has helped cities like New York to be more resilient in times of crisis.

The Internet

Not surprisingly, the Internet was one of the few telecommunications networks that was not seriously disrupted by the terrorist attacks. Designed around Cold War ideas for attack-proofing communications networks and built by a wide variety of competing firms, the Internet's present architecture is a dense web of overlapping, redundant, and inter-connected networks. According to Matrix.net, the leading analysts of traffic patterns on the Internet, network disruptions caused by the September 11 attacks were few and highly localized in the New York region.[2]

The ability of the Internet to withstand damage stems from its reliance on two protocols for routing traffic around damaged areas. The first is TCP/IP. TCP/IP's packetized nature means that data doesn't care how it gets to its destination, and thus can reroute itself as needed.

The second, less well-known cousin of TCP/IP is the Border Gateway Protocol (BGP). Because it is fairly common for a packet to transit several networks between its origin and destination, a system needs to be in place to determine the shortest path along each multi-network journey. This is complicated by the fact that each network has multiple entry and exit points in multiple cities. BGP provides a way for each network to "advertise" its shortest routes, while revealing nothing about its internal structure (which is a competitive secret). The advantage of BGP is that it provides for dynamic reassignment of routes, which comes in handy during a disaster like September 11. While it does take time (and often some human intervention) for networks to adjust, BGP is remarkably effective at keeping localized disruptions from propagating throughout the network. This is in contrast to other tightly coupled networks like the air transport system, in which localized problems tend to propagate quickly throughout the network (i.e. your flight from Los Angeles to New York is delayed because of bad weather in the Midwest).

These protocols helped the US Internet remain remarkably resilient while facing the challenges posed by September 11, 2001 which fell into two categories.

The first was the surge in traffic caused by a major media event. Globally, this surge was barely noticed and most major ISPs reported a five percent increase in data traffic across the board, which was easily handled by existing capacity. However, in the New York region late-night Internet activity caused significant congestion from around 9pm September 11 until early the next morning. The congestion on the New York area Internet lasted several days, according to John Quarterman of Matrix.net.

The second category of challenges had to do with physical damage and outages of Internet routing equipment. Intermittent power problems at two major Internet switching stations in New York did produce some routing problems for US-European Internet connections. However, BGP was able to quickly and effectively route around these problems. Many carriers such as Sprint and Qwest, offered each other capacity on transatlantic lines to deal with these outages.

Restoration of the Stock Exchange: Multiple Telecommunications Networks

Despite the extensive damage to lower Manhattan's telecommunications networks, and the resulting service disruptions that continued to ripple throughout these systems in the days following the attacks, restoring telecommunications and reopening the financial markets was of the utmost urgency. Furthermore, with transportation systems disrupted and firms displaced from the destroyed buildings, digital networks were called upon to provide flexibility for employers and employees as they tried to rebuild the city's economy.

The New York Stock Exchange managed to reopen just six days after the attacks, but it took an enormous effort on the part of telecommunications carriers and the power authority, Con Edison, to pull it off. By the start of trading on the Monday morning following the attacks, some 14,000 of the NYSE's 15,000 lines were back in service. The market is also dependent on its member firms, and Verizon managed to restore service to many of the financial companies that had been served out of 140 West Street, with as many as 2 million of the 3.5 million lost lines back up by the start of trading Monday morning (Young and Solomon 2001).

To accomplish this feat, Verizon focused almost exclusively on restoring service to the exchange and the largest financial firms for the first week after the disaster. Some 3,000 technicians and managers descended on lower Manhattan and Verizon trucks were parked over nearly every manhole south of Canal Street in lower Manhattan. Eighteen new SONET rings were installed, and one urban legend claims that Verizon employees sneaked past the Secret Service perimeter around Ground Zero to retrieve vital equipment from 140 West Street that was used to restore service to the stock exchange. Lines were laid on the street surface throughout the Financial District and covered with asphalt, in semi-permanent installations that would only be reburied underground weeks or even months later.

Other telecommunications companies rushed in to help restore services as well. AT&T's Network Disaster Recovery team was in place by midnight on September 12, less than 48 hours after the attacks (AT&T 2003).

Wireless Networks

Wireless carriers responded even more quickly, drawing upon lessons learned in previous urban disasters like the 1994 Northridge earthquake in Los Angeles. Using trailer mounted mobile cell sites, various carriers were quickly able to restore service to locations where antenna installations had been destroyed, such as Ground Zero. Other cell sites, whose landline connections had been lost due to the damage of wired networks near the WTC, were linked back to the grid using temporary point-to-point microwave links. These "wireless backhauls" patched into receivers along the Brooklyn waterfront and on tall buildings in midtown to reconnect the stranded cell sites to the citywide phone grid.

Even the network of a bankrupt wireless company took part in helping out in the rescue and recovery at Ground Zero. Metricom, a wireless data provider that had failed during the summer of 2001 had deployed a medium-speed (128 kbps) wireless data network throughout much of Manhattan. Due to financial problems the company had shut down its network on August 8, let go much of its staff, and was in bankruptcy proceedings at the time of the September 11 attacks. With the massive amount of evidence collection, and rescue-related communications needed at Ground Zero, New York City's Department of Information Technology and Telecommunications seized an opportunity, and petitioned the bankruptcy court to release some of the company's lower Manhattan assets (Charny 2001). Within two weeks, between 500 and 1,000 rescue workers were using the network to log evidence, communicate, and access death certificate records.

While digital networks were healing, both by themselves and with the help of their human caretakers, the city's social and economic networks were shaking off the shock of September 11 and beginning the long road back to normal life. In the process, they employed digital network technologies to support an initial restoration of normalcy, and eventually long-term recovery.

Telecommuting

Telecommuting played a major role in the region's ability to cope with the physical destruction of transportation links, and the closing of important river crossings like the Holland and Brooklyn-Battery Tunnels.

In the early 1980s, when the rise of the PC first opened the possibility of mass telecommuting, many futurists envisioned a highly decentralized society that could still provide access to the benefits of urban civilization; as Alvin Toffler called it, "the electronic cottage." Telework, or telecommuting, was seen as a way to eliminate travel caused by the spatial separation of nice places to live and places to work. Throughout that decade and into the 1990s, telecommuting made gains as both personal computers and high-speed data communications got cheaper and were widely deployed in homes around the US, Europe, and Asia. But unlike the wholesale decentralization envisioned by the futurists, home-based telecommuting remained a limited phenomenon. In fact, the knowledge-based workers most prone to telecommuting were the same workers that employers wanted to keep in close proximity to management and teammates, in order to maximize productivity.

Ironically, the banks of lower Manhattan had both the best communications technology at their disposal, and the greatest need to bring teams together face to face with each other, clients, and competitors to compete effectively. Partly as a result of this paradox, telecommuting never really caught on in the way it was originally envisioned – as a replacement for the traditional office worker's commute. Yet it has fulfilled what is arguably a far more important role, as a backup technology following natural disasters and civil emergencies. Some 3,000 employees of Lehman Brothers telecommuted on a daily basis during the second half of September while temporary offices were being procured.

Prior to September 11, natural disasters had provided the main impetus for the use of telecommuting as a component in disaster planning. The 1994 Northridge earthquake in Los Angeles is the archetypical example, because it destroyed so much transportation infrastructure. Yet it also coincided with the widespread arrival of two essential tools for telecommuting, personal computers and modems.

Telecommuting during emergencies in not limited to the United States alone. In 1998 Auckland, New Zealand experienced a cascading power failure that compromised all four of the central business district's redundant power arteries, causing a month-long blackout in February and March. Yet while the entire central business district was darkened – even the New Zealand Stock

Exchange had to shut down and switch to a backup center – it was widely reported that business went on as usual with workers setting up shop at home, pubs, and coffee shops relying on mobile phones, laptops, and modems to digitally reconstruct their social and professional networks.

Relocation of Displaced Firms

The destruction of the World Trade Center illustrated the fragility of corporate office buildings. However, until some strategy can be found to eliminate the need for office buildings, firms must house their workers somewhere. In the wake of September 11, the need for quick return to the markets, the scale of destruction, and the shortage of readily available space forced many firms to use digital network technologies to help find creative solutions.

Perhaps the most widely used strategy was the activation of backup sites by many financial firms. Backup "hot sites" tend to be located within the same region as primary sites for both human and technological reasons. Firstly, at least some of the workers who are expected to staff backup sites would be those displaced from the main worksite. Secondly, there are limits to how far digital network infrastructure can be stretched to decentralize corporate information systems. In the financial services sector in particular, the amount of data that needs to be archived and stored at remote disaster recovery facilities stretches the capabilities of even the highest-speed networks. The precise timing required by these ultra-high capacity networks limits the distance across which such systems can function. For example, IBM's ESCON protocol used by mainframes to talk to remote mass storage devices is limited to about 20 miles.

Long-term Recovery

Finally, digital networks are critical to the long-term recovery of the New York region from the September 11 attacks. Furthermore, a number of precautions to "harden" digital network infrastructure need to be considered. In the long run, one of the major outcomes of September 11 may be to firmly place the planning, provisioning, and preparedness of digital network infrastructure solidly on the agenda of urban planners and policymakers. A number of key issues and strategies have come to the forefront.

Redundancy Through Wireless Alternatives

The destruction of large concentrations of wired network infrastructure around the World Trade Center site had devastating consequences on telecommunications services in lower Manhattan and the entire New York region. One strategy which was widely used to quickly restore communications links was the deployment of temporary point-to-point wireless links. Microwave links were widely used to connect rooftop sites to uplinks in midtown Manhattan, New Jersey, and Brooklyn. Laser-based optical networks were deployed by AT&T to provide high bandwidth backbone links to New Jersey.

A recent report by the Lower Manhattan Telecommunications Users' Working Group recommends the development of a rooftop wireless network in Lower Manhattan to provide a fallback in future disasters. At an estimated cost of less than US$1 million, a 6 node rooftop mesh network could serve over 2,000 buildings in the Financial District.

Terrorism, Threat and Urban Decentralization

Many urban observers have argued that the post-September 11 city will be increasingly decentralized, as firms seek to spread risk and reduce the number of large targets they must insure. As Lehman Brothers' Managing Director David Shulman remarked last fall:

> ...the impact of imposing a terrorism tax on the economy that is undoing many of the agglomeration economies now extant in the economy. Put bluntly, the gains associated with the clustering of functions within a firm and of firms in similar industries are being offset by the risk of terrorism (Lyne 2001).

Preliminary evidence from New York City suggests that there has been little to support this point of view. In fact, according to calculations by New York University's Real Estate Institute, some 80 percent of displaced demand from the World Trade Center and surrounding buildings was accommodated in existing office space throughout the five boroughs. The remaining 20 percent, mostly in the urban New Jersey waterfront just across from lower Manhattan, have not committed to long-term leases. Put simply, the "terrorism tax" does not appear to be strong enough to overcome the advantages of agglomeration in a large business cluster like the Financial District. In fact, some of the most affected firms have made even more substantial commitments to New York. Lehman Brothers, for example, purchased a one million sq ft office building in midtown Manhattan.

The Internet Needs Power: the Web of Urban Networks

For their resiliency, digital networks remain highly dependent upon older, less flexible infrastructure systems for food and shelter. Much of lower Manhattan's telecommunications would have remained online if Con Edison had been able to maintain the electricity grid. The lack of physical access to Manhattan also means that many telecommunications networks must share the same physical conduits into the Manhattan central business district – the Lincoln, Holland, and Battery Tunnels. Should these transportation arteries be compromised by a terrorist plot (one had been uncovered by Federal investigators in the mid-1990s), Manhattan might be cut off from Europe or lose much of its ability to talk to the rest of the United States.

Targeting Digital Networks

While the Internet's performance was highly regarded by many, it must be remembered that the Internet was not targeted by the September 11 attacks. The

Nimda virus (which was determined to be unrelated to 9/11) which struck some 85,000 Internet servers a week later on September 18, caused enormous network congestion in cities worldwide at an estimated economic cost of US$1.2 billion – placing it on the scale of the WTC attack (Computer Economics 2003).

The Internet does appear vulnerable to physical attack as well, especially in dense urban areas. And it is also unclear just how meshed the Internet architecture actually is, particularly at the global scale.[3] The New York metropolitan area is the world's largest single international bandwidth hub, according to market analyst Telegeography, Inc. Yet the three main Internet interconnection points in the region lie within miles of each along the west side of lower Manhattan. There has been speculation that taking out three carrier hotels in Manhattan (111 8th Avenue, 25 Broadway, and 60 Hudson Street) would effectively separate the US Internet from Europe for several days or more.

Telecommuting as a Backup in Emergencies

Post-disaster telecommuting experiences had long-term impacts on the evolution of the Los Angeles region. Northridge was a big push for many firms who had ignored telecommuting to reconsider its possibilities for reducing long-distance commuting, which it turn had long term benefits for the region. As one observer put it:

> [T]he January 1994 earthquake provided the jolt that many California companies needed to try new, innovative work methods to change their corporate cultures. Now the broader use of flextime, telecommuting, compressed work-weeks, vanpools and carpools are subtly changing the corporate culture for the better… (Dutton 1994).

Pacific Bell's telecommuting information hotline, set up in the quake's wake, fielded some 2,000 calls from small businesses seeking to use telecommuting as a solution to managing transportation problems.

The long-term impacts of Northridge on the LA basin have been to accelerate the decentralization of the region's already highly dispersed commercial infrastructure. Because of the role telecommuting played in coping with the disaster, it has become an integral part of the region's ongoing battle to provide mobility and accessibility. Perhaps the most telling observation of the technological response to the quake appeared in *PC World*, where one group of editors wrote "It took an earthquake to make Los Angelenos avoid their cars and consider telecommuting … [but] government officials pledged to open freeways by year end, so it remains to be seen whether Angelenos abandon their sacred autos for modems" (Glitman *et al.* 1999).

Notes

1 According to a Pew Research Center study, only one percent of people surveyed first found out about the attacks via the Internet, compared to 44 percent by television and 15 percent by telephone.

2 Ironically, these disruptions were caused by power failures at two network exchange points in Lower Manhattan in the week following the attacks, illustrating the inter-dependency of urban technological infrastructures (see Graham and Marvin (2001) *Splintering Urbanism: Networked Infrastructures, Technological Mobilities and the Urban Condition*, London: Routledge).

3 Consider the case of South Africa. All of that nation's largest connections to the global Internet backbone are routed through New York City. As a result of the Internet outages in New York City, South Africa lost about 50 percent of its international Internet capacity for a week (National Research Council 2003).

References

"AT&T Network Disaster Recovery Team – Deployments," Online, Available at: http://www.att.com/ndr/deployment.html (accessed December 5, 2003).

Charny, B. (2001) "Ricochet rebounds at Ground Zero," Cnet News, Online, Available at: http://news.com.com/2100-1033-273743.html?tag=bplst (accessed December 5, 2003).

Computer Economics, Inc. Carlsbad, CA http://www.computereconomics.com/.

Cone, E. and Gallagher, S. (2001) "Cantor-Fitzgerald: 47 Hours," Baseline, Online, Available at: http://www.baselinemag.com/article2/0,3959,36807,00.asp (accessed December 5, 2003).

Dutton, G. (J1994) "Can California Change Its Corporate Culture?" *Management Review*, Vol. 83, No. 6, pp. 49–54.

Dwyer, J., Flynn, K. and Fessenden, F. (2002) "9/11 Exposed Deadly Flaws in Rescue Plan," *New York Times*, July 7, p. 1.

Gaudin, S. (2001) "Lehman Brothers' network survives," *NetworkWorldFusion*, Online, Available at: http://www.nwfusion.com/research/2001/1126feat.html (accessed December 5, 2003).

Glitman, R., Abernathy, J., Bertolucci, J., Brownstein, M. *et al.* (1994) "Temblor telecommuters," *PC World*, Vol. 12, No. 4, pp.63–4.

Graham, S. (2001) "In a Moment: On Global Mobilities and the Terrorised City," Circulated by email, September 2001.

Graham, S. and Marvin, S. (2001) *Splintering Urbanism: Networked Infrastructures, Technological Mobilities and the Urban Condition*, London: Routledge.

Lyne, J. (2001) "Real estate strategy, Post 09/11/01: In Part, a Whole New Ballgame," Site Selection.

National Research Council of the National Academies (NRC) (2003) *The Internet Under Crisis Conditions: Learning From September 11*, Washington, D.C.: National Academies Press, Computer Science and Telecommunications Board.

Ohlson, K. (2001) "Businesses start the recovery process," *NetworkWorldFusion*, Online, Available at: http://www.nwfusion.com/news/2001/0912rec.html (accessed December 5, 2003).

Young, S. and Solomon, D. (2001) "Verizon Effectively Rebuilds Network for NYSE," *Wall Street Journal*, September 18, p. B7.

PART III
Crosscutting Themes

CHAPTER NINE

Economic Impacts: The Growing Use of Information Technology in Infrastructure Industries

David Henry and Jesus Dumagan[1]

Introduction

Between 1996 and 2001, infrastructure industries, which together accounted for 8–9 percent of the nation's Gross Domestic Product (GDP), owned about one-third of all capital stock in information technology (IT) equipment and software. But despite this heavy investment in digital products, in general, the productivity growth in infrastructure industries was not dramatic and did not contribute significantly to the overall labor productivity acceleration that happened in the late 1990s.[2]

The following discussion and quantitative assessment of infrastructure industries does not answer why these industries have a lack of labor productivity responsiveness, but it does demonstrate the central importance of digital products and services to our nation's infrastructure.

The economic performance of traditional infrastructure industries – those in the utilities, transportation, and communications sectors of the economy – are increasingly dependent on a *digital* support structure, i.e. on the goods and services produced by IT industries. IT industries support not only the traditional infrastructure industries but practically all industries in the United States.

IT industries[3] comprise those producing IT hardware such as computers, semiconductors, electrical and mechanical measuring devices, and communication equipment. They also encompass industries that support the use and care of digital equipment. They are communications services, computer and data processing services, and software industries.

In this chapter, we:

- examine developments in IT industries and their impact on the US economy

- examine the economic performance of infrastructure industries and explore the premise that a digital infrastructure supports them
- compare the relative level of IT intensity of infrastructure industries with all other (non-farm) industries and then examine their contribution to overall labor productivity growth
- examine how IT technologies in infrastructure industries are displacing some jobs and explore how job displacement is affecting labor productivity.

IT-Producing Industries: Dramatic Growth Leads to Significant Impact on the Economy

As other chapters in this book have demonstrated, IT equipment and software have become an integral component of the production efficiency and capability of infrastructure industries. Business purchases of IT equipment and software led to the recent acceleration in labor productivity growth, in infrastructure industries during 1999 and 2000 and in all other US industries over the 1999–2001 period (see Figure 9.1 for the list of IT industries used in this analysis).[4]

Hardware	**Software/Computer Services**
Computers and equipment	Computer programming
Wholesale trade of computers and equipment	Prepackaged software
Retail trade of computers and equipment	Wholesale trade of software
Calculating and office machines	Retail trade of software
Magnetic and optical recording media	Computer-integrated system design
Electron tubes	Computer processing data preparation
Printed circuit boards	Information retrieval services
Semiconductors	Computer services management
Passive electronic components	Computer rental and leasing
Industrial instruments for measurement	Computer maintenance and repair
Instruments for measuring electricity	Computer related services, nec
Laboratory analytical instruments	
Communications Equipment	**Communications Services**
Household audio and video equipment	Telephone and telegraph communications
Telephone and telegraph equipment	Cable and other pay TV services
Radio and TV communications equipment	

9.1 *Information Technology (IT) Producing Industries.*
Source: Economics and Statistics Administration, US Department of Commerce (2003) *Digital Economy 2003*, 1997 North American Industrial Classifications System industry codes available from http://www.esa.doc.gov

Demand Grew Rapidly for IT Goods and Services

Businesses, in the latter part of the 1990s, made a clear decision to accelerate spending on IT equipment and software in order to cash in on the improvements in production, productivity, and profitability brought on by these IT products. Before the emergence of the Internet in 1995, the annual growth in business investments for IT equipment (Hardware and Communications equipment) and software averaged nine percent (1991–1995) in current dollars.[5] In 1995, business spending on IT equipment and software was US$263 billion. Between 1996 and 2000, all business spending (capital investing) for IT equipment and software rose by an average annual 12 percent, in current dollars, to US$468 billion.

In 2001, the growth spurt in spending for IT came to an abrupt halt as the economy slowed from a four to five percent average annual growth rate between 1997 and 2000 to only 0.1 percent. Business spending on IT equipment and software dropped for the first time in a decade, by ten percent in nominal dollars and six percent in real dollars.

While business spending for IT capital equipment is the most significant market for IT producers, it is not the only market. Businesses also spend significant sums for communications and computer services. This spending is not considered to be investment. In 2000, we estimate that businesses spent US$258 billion on communications services and another US$262 billion on computer services. Personal consumption and government spending are other important markets for IT. Consumers spent US$165 billion on computers and computer equipment and communications services in 2000. The Federal, state, and local governments spent about US$20 billion on IT goods and services.

IT Industries Respond with Dramatic Production Growth

Table 9.1 provides the current and constant dollar value of GDP for IT producing industries, 1996 through 2001. Current dollar GDP by industry is the contribution of each private industry to the nation's GDP. It is the industry's value added, which is equal to its gross output (which consists of sales, receipts, or revenue) minus the costs of production (intermediate inputs).[6]

Between 1997 and 2000, GDP by industry of IT producing industries rose by an annual average ten percent, current dollars, and 18 percent, constant (1996) chained dollars. In 2001, these industries declined by six percent, current dollars, and increased by one percent in constant dollars. In 2001, moderate growth continued in the software, computer services and the communications services sectors, while most of IT hardware and communications equipment stalled or declined sharply.

When adjusted for inflation, GDP for the IT producing industries reflected a different mix of growth rates.[7] Prices of computers fell by about 24 percent per year, 1997 through 2000, then 16 percent in 2001. Semiconductor prices fell an average 15 percent per year followed by nine percent in 2001. Reflecting these dramatic price declines in IT hardware and communications equipment, real growth in these sectors increased sharply as compared to growth in IT services

Table 9.1 *Gross Domestic Product (GDP) of US IT-producing Industries*

	1996	*1997*	*1998*	*1999*	*2000*	*2001*
IT-producing industries	Nominal GDP (billion dollars)					
Computers and equipment, calc. machines	51.2	56.1	66.7	62.3	55.0	46.4
Computers and equipment wholesale sales	59.8	68.2	72.4	77.2	72.2	58.8
Computer and equipment retail sales	4.8	5.5	6.1	6.4	6.4	5.6
Electron tubes	1.6	1.6	1.6	1.8	1.4	1.2
Printed circuit boards	5.0	5.1	5.0	5.6	6.6	4.7
Semiconductors	51.6	64.1	60.0	68.0	67.9	44.1
Passive electronic components	15.3	13.0	12.4	12.7	14.8	11.0
Industrial instruments for measurement	3.1	4.9	4.8	4.6	4.8	4.6
Instruments for measuring electricity	5.7	9.0	8.8	9.0	10.3	8.2
Laboratory analytical instruments	3.0	4.1	4.5	4.7	4.7	4.7
Computer programming services	29.4	35.0	47.7	56.4	63.3	61.1
Prepackaged software	47.7	52.3	61.4	69.1	76.0	77.7
Prepackaged software wholesale sales	3.5	3.8	4.1	4.3	4.0	3.3
Prepackaged software retail sales	0.6	0.7	0.8	0.8	0.8	0.7
Computer integrated system design	34.0	43.9	58.2	66.1	75.1	73.7
Computer processing, data preparation	23.1	24.4	26.3	30.0	33.6	37.9
Information retrieval services	5.3	6.9	9.7	16.4	25.3	26.2
Computer services management	10.2	11.8	9.3	11.9	12.8	14.9
Office machinery rental and leasing	4.2	4.8	6.5	7.3	7.8	6.6
Computer maintenance and repair	5.9	6.6	8.0	8.5	8.5	8.7
Computer related services, nec	2.5	3.4	6.1	7.5	9.3	9.5
Household audio and video equipment	1.7	2.4	2.7	2.9	3.2	3.1
Telephone equipment, exc. ext. modems	17.0	23.8	25.4	32.7	35.8	28.0
Radio and TV communications equipment	18.0	24.3	22.8	23.1	26.8	22.8
Magnetic and optical and recording media	2.3	2.3	2.2	1.9	1.4	1.0
Telephone and telegraph communications	163.9	166.7	173.9	193.7	208.0	218.5
Cable and other pay TV services	18.7	21.5	32.2	36.9	41.8	45.8
Rest-of-US economy	7,191.3	7,622.5	8,073.0	8,491.4	9,075.3	9,370.5
Total nominal gross domestic income (GDI)	7,780.3	8,288.6	8,812.5	9,313.1	9,953.1	10,199.4
Share of IT-producing industries in total GDI (percent)	7.6	8.0	8.4	8.8	8.8	8.1

Source: Authors' calculations based on Census data and GDP-by-industry series and US Gross Domestic Income (GDI) from BEA using BEA's chained-dollar aggregation procedure. Because of chained-dollar non-additivity, aggregate real GDI is not equal to the arithmetic sum of the real GDI components.

Note

"Chained" dollars are real values based on a chain-type quantity index-devised by the US Bureau of Economic Analysis (BEA) – that is tied to the Fisher ideal quantity index formula. The latter formula by definition, uses prices as weights. However, to generate chained dollars, the prices do not have a fixed base period. Instead, the quantity index uses as weights the prices from two adjoining periods defined by the index. So, the weights change as time moves to impart a "chaining effect. (See Landefeld and Parker 1997).

Table 9.1 *Continued*

	1996	*1997*	*1998*	*1999*	*2000*	*2001*
IT-producing industries	Real GDP (billion chained 1996 dollars)					
Computers and equipment, calc. machines	51.2	72.6	117.0	142.8	145.2	150.0
Computers and equipment wholesale sales	59.8	88.2	127.0	177.1	190.7	190.1
Computer and equipment retail sales	4.8	5.5	6.2	6.5	6.6	5.7
Electron tubes	1.6	1.7	2.1	2.7	2.6	2.8
Printed circuit boards	5.0	5.6	6.7	8.6	12.8	11.1
Semiconductors	51.6	70.4	79.4	105.2	130.7	103.3
Passive electronic components	15.3	14.3	16.4	19.6	28.4	25.9
Industrial instruments for measurement	3.1	4.9	4.7	4.5	4.7	4.5
Instruments for measuring electricity	5.7	9.0	8.7	8.9	10.2	8.0
Laboratory analytical instruments	3.0	4.1	4.5	4.7	4.7	4.6
Computer programming services	29.4	34.4	46.2	53.5	58.1	56.2
Prepackaged software	47.7	53.4	64.5	72.0	78.1	78.4
Prepackaged software wholesale sales	3.5	3.9	4.3	4.5	4.2	3.3
Prepackaged software retail sales	0.6	0.7	0.8	0.9	0.9	0.7
Computer integrated system design	34.0	43.2	56.4	62.7	69.0	67.8
Computer processing, data preparation	23.1	24.0	25.5	28.5	30.9	34.8
Information retrieval services	5.3	6.8	9.4	15.5	23.2	24.0
Computer services management	10.2	11.6	9.0	11.3	11.7	13.7
Office machinery rental and leasing	4.2	4.7	6.3	6.9	7.2	6.1
Computer maintenance and repair	5.9	6.5	7.7	8.0	7.8	8.0
Computer related services, nec	2.5	3.3	5.9	7.1	8.6	8.8
Household audio and video equipment	1.7	2.4	2.8	3.1	3.6	3.7
Telephone equipment, exc. ext. modems	17.0	24.0	26.8	35.8	40.6	32.7
Radio and TV communications equipment	18.0	24.5	24.1	25.3	30.3	26.6
Magnetic and optical and recording media	2.3	2.3	2.4	2.1	1.6	1.2
Telephone and telegraph communications	163.9	167.9	181.3	205.3	236.7	265.7
Cable and other pay TV services	18.7	19.8	27.4	29.9	31.4	37.7
Rest-of-US economy	7,191.3	7,427.0	7,701.1	7,927.0	8,239.5	8,242.2
Aggregate real gross domestic income (GDI)*	7,780.3	8,130.1	8,538.8	8,895.8	9,311.4	9,321.3
Growth of aggregate real GDI (percent)		4.50	5.03	4.18	4.67	0.11

Note

By definition, nominal GDP = nominal GDI + statistical discrepancy. GDP and GDI are alternative measures of the US economy's size. GDP is based on expenditure-side components while GDI is based on income or product-side components. Because the two sides use different data sources, the statistical discrepancy serves as a "balancing item." In recent years, the statistical discrepancy has been negative because GDP is less than GDI. For example, the statistical discrepancy amounted to −117.3 billion dollars in 2003.

(software and computer services and communications services). Between 1997 and 2000, real growth in IT hardware increased by an average annual 27 percent. IT communications equipment increased by an annual average 19 percent.

The growth in demand for IT goods and services and the resulting output growth in IT industries was responsible for a significant portion of the four to five percent real growth rate in the US economy (total GDP) between 1997 and 2000. Between 1997 and 2000, the growth in IT industries was equal to about one-third of total US GDP growth (see Table 9.2) even though, as an aggregate sector, these industries accounted for between seven to nine percent of GDP.

The flip side of IT's contribution to real growth is IT's contribution to keeping overall inflation low. Between 1997 and 2001, inflation, as measured by the implicit GDP deflator, averaged less than two percent per year. IT producing industries contributed an average −0.6 percentage point to inflation growth while all other industries in the economy averaged 2.4 percentage points. Overall, average US inflation over the period was 1.8 percent per year (see Table 9.3).

US IT Producing Industries Lead IT Producers in Other Countries

US IT producing industries have been leaders in production of IT goods and services compared to their equivalent industries in other countries and all US industries have been more successful in incorporating the IT technological efficiencies resulting in improved labor productivity. On the other hand, production of IT goods and services in foreign countries has gained in size and in importance to their respective economies.

The Organisation for Economic Co-operation and Development (OECD) has, over the last several years, provided country comparisons of IT.[8] This organization (which represents some 30 countries, including the US) finds that:

- By the end of the growth cycle, in 2000, US firms continued to invest more heavily in IT equipment and software than all other OECD countries. US firms invested 38 percent of their non-residential business investment dollars in IT equipment and software. Businesses in Finland invested 30 percent; the U.K. and Canada 22 percent; Germany, 19 percent; Italy and Japan, 17 percent; and France, 13 percent.

- In 2001, the production by US IT producing industries accounted for about 11 percent of the GDP represented by non-farm business. Finland's IT producing industries accounted for 16 percent of non-farm business. Japan's IT industries accounted for ten percent; Canada and France eight percent; Italy, seven percent; and Germany, six percent.

Table 9.2 Contribution of IT-producing Industries to Economic Growth

	1997	*1998*	*1999*	*2000*	*2001*
	Percentage points				
Computers and equipment, calc. machines	0.242	0.358	0.146	0.010	0.016
Computers and equipment wholesale sales	0.321	0.313	0.284	0.059	−0.002
Computer and equipment retail sales	0.009	0.008	0.004	0.000	−0.008
Electron tubes	0.002	0.004	0.005	−0.001	0.001
Printed circuit boards	0.006	0.011	0.015	0.026	−0.008
Semiconductors	0.229	0.090	0.203	0.158	−0.129
Passive electronic components	−0.013	0.021	0.025	0.055	−0.012
Industrial instruments for measurement	0.022	0.002	−0.002	0.002	−0.003
Instruments for measuring electricity	0.042	−0.003	0.003	0.014	−0.022
Laboratory analytical instruments	0.013	0.004	0.003	0.000	−0.001
Computer programming services	0.064	0.144	0.086	0.053	−0.020
Prepackaged software	0.072	0.128	0.081	0.063	0.003
Prepackaged software wholesale sales	0.005	0.004	0.003	−0.004	−0.008
Prepackaged software retail sales	0.001	0.001	0.000	0.000	−0.001
Computer integrated system design	0.118	0.162	0.073	0.072	−0.013
Computer processing, data preparation	0.012	0.018	0.035	0.027	0.043
Information retrieval services	0.020	0.032	0.071	0.088	0.009
Computer services management	0.019	−0.032	0.027	0.005	0.021
Office machinery rental and leasing	0.008	0.019	0.007	0.003	−0.012
Computer maintenance and repair	0.008	0.015	0.004	−0.003	0.002
Computer related services, nec	0.011	0.031	0.014	0.017	0.002
Household audio and video equipment	0.008	0.005	0.003	0.005	0.000
Telephone equipment, exc. ext. modems	0.089	0.032	0.095	0.046	−0.068
Radio and TV communications equipment	0.083	−0.005	0.013	0.048	−0.032
Magnetic and optical and recording media	0.000	0.000	−0.003	−0.004	−0.004
Telephone and telegraph communications	0.051	0.157	0.257	0.304	0.245
Cable and other pay TV services	0.015	0.103	0.033	0.021	0.080
Economic growth contributions of IT-producing industries (A)	1.46	1.62	1.49	1.07	0.08
Economic growth contribution of the rest-of-US economy (B)	3.04	3.41	2.70	3.61	0.03
	Percent				
Economic growth = Growth of aggregate real GDI (A + B)[a]	4.50	5.03	4.18	4.67	0.11
Economic growth share of IT-producing industries [(A/(A + B)) × 100]	32.4	32.2	35.5	22.8	71.7

Source: Authors' calculations based on Census and BEA data and on BEA's exact formula for component contributions to chained-dollar aggregate growth.

Note
a See the last row of Table 9.1.

Table 9.3 *Contribution of IT-Producing Industries to Price Growth*

	1997	*1998*	*1999*	*2000*	*2001*
	Percentage points				
Computers and equipment, calc. machines	0.242	0.358	0.146	0.010	0.016
Computers and equipment, calc. machines	−0.175	−0.226	−0.193	−0.086	−0.103
Computers and equipment wholesale sales	−0.210	−0.257	−0.226	−0.110	−0.133
Computer and equipment retail sales	0.000	0.000	0.000	0.000	0.000
Electron tubes	−0.002	−0.004	−0.003	−0.004	−0.003
Printed circuit boards	−0.006	−0.011	−0.009	−0.014	−0.011
Semiconductors	−0.068	−0.137	−0.111	−0.157	−0.109
Passive electronic components	−0.017	−0.028	−0.022	−0.032	−0.025
Industrial instruments for measurement	0.000	0.000	0.000	0.000	0.001
Instruments for measuring electricity	0.000	0.000	0.000	0.000	0.001
Laboratory analytical instruments	0.000	0.000	0.000	0.000	0.001
Computer programming services	0.006	0.008	0.013	0.020	−0.001
Prepackaged software	−0.013	−0.018	0.006	0.010	0.014
Prepackaged software wholesale sales	−0.001	−0.001	0.000	0.001	0.001
Prepackaged software retail sales	0.000	0.000	0.000	0.000	0.000
Computer integrated system design	0.008	0.009	0.015	0.023	−0.001
Computer processing, data preparation	0.005	0.005	0.007	0.011	0.000
Information retrieval services	0.001	0.002	0.003	0.007	0.000
Computer services management	0.002	0.002	0.003	0.004	0.000
Office machinery rental and leasing	0.001	0.001	0.002	0.003	0.000
Computer maintenance and repair	0.001	0.001	0.002	0.003	0.000
Computer related services, nec	0.001	0.001	0.002	0.003	0.000
Household audio and video equipment	0.000	−0.001	−0.001	−0.001	−0.001
Telephone equipment, exc. ext. modems	−0.003	−0.012	−0.012	−0.013	−0.010
Radio and TV communications equipment	−0.003	−0.012	−0.009	−0.009	−0.008
Magnetic and optical and recording media	0.000	−0.001	−0.001	−0.001	0.000
Telephone and telegraph communications	−0.015	−0.069	−0.034	−0.150	−0.142
Cable and other pay TV services	0.020	0.025	0.019	0.031	−0.040
Price growth contributions of IT-producing industries (A)	−0.47	−0.72	−0.55	−0.46	−0.57
Price growth contribution of the rest-of-US economy (B)	2.42	1.96	1.99	2.56	2.93
	Percent				
Price growth = Growth of GDI implicit price deflator (A + B)[a]	1.95	1.23	1.44	2.10	2.36
Price growth share of IT-producing industries [(A/(A + B)) × 100]	−24.0	−58.7	−38.2	−22.0	−24.1

Source: Authors' calculations based on BEA's exact formula for component contributions to chained-dollar aggregate growth.

Note
a The GDI implicit price deflator is the ratio of the total nominal GDI to the aggregate real GDI in Table 9.1.

Infrastructure Industries: Trends in Growth, Prices, Capital Stock and IT Investment

Slower Growth and Rising Prices

While IT producing industries were characterized by rapidly increasing production and drastically falling prices during the 1997 to 2001 period, infrastructure industries were characterized by slower, but steadier, growth and rising prices.

Infrastructure industries included in this analysis include utilities (Electric power generation, transmission and distribution, Natural gas distribution, and Water, sewage, and other systems), transportation industries (Air, Water, Truck, Transit and urban ground transportation, Pipelines, and Transportation services), and communications (Telephone and telegraph communications and Radio and TV broadcasting). (Figure 9.2 lists infrastructure industries used in this analysis by their 1997 North American Industry Classification code.) The Telephone and telegraph communications industry is considered both an infrastructure industry as well as an IT producing industry.

In contrast to the ten percent average annual growth (nominal value of GDP by industry) in IT producing industries between 1997 and 2000, infrastructure industries grew an average five percent (see Table 9.4).

Overall spending across the entire economy on services from infrastructure industries trended down as the economy became less economically dependent on utilities, transportation, and communications during 1996 through 2000. In 1996, infrastructure industries accounted for about 8.6 percent of GDP. By 2000, they were 8.1 percent. In 1987, these industries accounted for nine percent of the economy.

In inflation adjusted dollars, the contrast of growth between infrastructure industries and IT producing industries was even sharper. Since prices in IT

Industries	**NAICS***
Electric power generation, transmission and distribution	2211
Natural gas distribution	2212
Water, sewage, and other systems	2213
Air transportation	481
Rail transportation	482
Water transportation	483
Truck transportation	484
Transit and ground passenger transportation	485
Pipelines	486
Transportation services	87488
Telephone and telegraph communications	5133
Radio and TV broadcasting	5131

*1997 North American Industry Classification System (NAICS)

9.2 *Infrastructure Industries.*

Table 9.4 *Gross Domestic Product (GDP) of US Infrastructure Industries*

	1996	*1997*	*1998*	*1999*	*2000*	*2001*
Infrastructure Industries	Nominal GDP (billion dollars)					
Railroads	23.4	23.0	24.3	23.1	24.8	25.8
Local and interurban transportation	13.4	14.9	16.8	17.9	18.3	19.1
Trucking	92.1	99.4	114.1	121.4	123.3	126.0
Water transportation	12.2	13.1	13.6	13.4	15.0	15.7
Air transportation	70.8	78.6	85.8	90.0	91.9	80.2
Pipelines, exc. Natural gas	5.7	5.8	6.1	6.2	6.3	6.5
Transportation services	25.7	27.1	28.0	29.9	34.1	32.9
Telephone and telegraph	163.9	166.7	179.4	193.7	208.0	218.5
Radio and TV broadcasting	50.7	54.1	59.1	63.6	71.1	72.9
Electric, gas, and sanitary services	208.3	205.9	204.8	211.0	216.5	221.9
Rest-of-US economy	7,114.1	7,600.0	8,080.5	8,542.9	9,143.8	9,379.9
Total nominal gross domestic income (GDI)	7,780.3	8,288.6	8,812.5	9,313.1	9,953.1	10,199.4
Share of infrastructure industries in total GDI (percent)	8.56	8.31	8.31	8.27	8.13	8.03
Infrastructure Industries	Real GDP (billion chained 1996 dollars)					
Railroads	23.4	22.8	22.8	22.4	25.0	25.0
Local and interurban transportation	13.4	14.7	15.5	16.8	17.2	17.2
Trucking	92.1	90.5	95.5	100.2	103.4	99.3
Water transportation	12.2	13.2	13.2	12.0	13.4	13.3
Air transportation	70.8	75.2	76.8	80.7	84.8	78.3
Pipelines, exc. Natural gas	5.7	6.2	6.4	6.6	6.4	6.0
Transportation services	25.7	26.4	27.8	29.8	32.4	31.2
Telephone and telegraph	163.9	167.9	181.3	205.3	236.7	265.7
Radio and TV broadcasting	50.7	49.9	50.3	51.5	53.4	60.0
Electric, gas, and sanitary services	208.3	202.0	193.7	208.8	213.9	194.3
Rest-of-US economy	7,114.1	7,461.7	7,856.6	8,164.2	8,530.5	8,542.0
Aggregate real gross domestic income (GDI)[a]	7,780.3	8,130.2	8,538.9	8,896.1	9,311.6	9,321.6
Growth of aggregate real GDI (percent)		4.50	5.03	4.18	4.67	0.11

Source: Authors' calculations based on GDP-by-industry series and US GDI from BEA using BEA's chained-dollar aggregation procedure. Because of chained-dollar non-additivity, aggregate real GDI is not equal to the arithmetic sum of the real GDP components. Note that aggregate real GDI above is slightly different from that in Table 9.1 because the chained-dollar aggregation procedure is sensitive to the level of detail of the components. Notice that there are more IT-producing industries (Table 9.1) than there are infrastructure industries. However, the difference between the aggregate real GDI in Table 9.1 and that above is so small that the growth rates end up equal.

Note:
a By definition, nominal GDP = nominal GDI + statistical discrepancy. GDP and GDI are alternative measures of the US economy's size. GDP is based on expenditure-side components while GDI is based on income or product-side components. Because the two sides use different data sources, the statistical discrepancy serves as a "balancing item." In recent years, the statistical discrepancy has been negative because GDP is less than GDI. For example, the statistical discrepancy amounted to −117.32 billion dollars in 2001.

producing industries declined dramatically over the period, IT producing industries experienced a real average annual growth of 18 percent. On the other hand, infrastructure industries, in real terms, grew an average four percent per year between 1997 and 2000, below half of their nominal growth. This means that despite growth in the infrastructure sector, in general, industry prices rose over the period. Rising prices in this sector can be attributed to any number of factors that restrict price competition including relatively inelastic demand, regulatory burdens, and a dominance of a foreign supply of oil with fixed pricing. On the other hand, prices in Communications, both an IT producing industry and an infrastructure industry, declined over the period – an annual average 3.2 percent.

In 2001, when the economy stalled, spending associated with infrastructure industries also stalled. In particular, revenues for Air transportation and Transportation services declined sharply. These industries were also clearly affected by the September 11, 2001 terrorist incident. Together, Air transportation and Transportation services represented 13.8 percent of the output (GDP by industry) of infrastructure industries in 2001. In 2001, output in Air transportation declined by 13 percent (nominal) and output in Transportation services declined by four percent.

The largest of the infrastructure industries in 2001 was the production, generation, and distribution of utilities (electric, gas, and sanitary services) and the telephone and telegraph communications industries. Sanitary services industry includes water and sewage and water treatment facilities. Together, utilities and communications accounted for over half of the value of output of all infrastructure industries. In 2001, output in electric, gas, and sanitary services increased by two percent and output in communications increased by five percent. The group of industries representing the transportation sector accounted for the remaining value of output.

Between 1997 and 2000, while growth in the IT producing sector equated to a large share of GDP growth, growth in the infrastructure industries was much less (see Table 9.5).

In fact, without the contribution of the communications sector (which is also an IT producing industry), real overall GDP growth of the infrastructure industries would, in the 1996 through 2001 period, have been negative. While IT producing industries represented about eight percent of the economy and accounted for a third of its economic growth, infrastructure industries that also represented about eight percent of the economy between 1996 and 2000 accounted for only a twelfth of the overall economic growth.

In contrast to price effects of IT producing industries, price changes in infrastructure industries contributed to the overall inflation rate (see Table 9.6). In 1997, infrastructure industries contributed about 13 percent to the overall inflation rate of two percent (the implicit GDP deflator) in 1997. In 1998, these industries accounted for 27 percent of the 1 percent inflation and contributed five percent to the two percent inflation rate in 2001. Air transportation and Telephone and telegraph communications both helped to keep prices down.

Table 9.5 *Contribution of Infrastructure Industries to Economic Growth*

	1997	1998	1999	2000	2001
	Percentage points				
Railroads	−0.008	0.000	−0.005	0.028	0.000
Local and interurban transportation	0.017	0.010	0.016	0.005	0.000
Trucking	−0.021	0.069	0.064	0.041	−0.050
Water transportation	0.013	0.000	−0.015	0.017	−0.001
Air transportation	0.057	0.021	0.049	0.048	−0.068
Pipelines, exc. Natural gas	0.006	0.002	0.002	−0.002	−0.004
Transportation services	0.009	0.017	0.023	0.028	−0.013
Telephone and telegraph	0.051	0.159	0.261	0.304	0.245
Radio and TV broadcasting	−0.011	0.005	0.016	0.026	0.083
Electric, gas, and sanitary services	−0.081	−0.103	0.176	0.055	−0.209
Economic growth contributions of infrastructure industries (A)	0.03	0.18	0.59	0.55	−0.02
Economic growth contribution of the rest-of-US economy (B)	4.46	4.85	3.59	4.12	0.12
	Percent				
Economic growth = Growth of aggregate real GDI (A + B)[a]	4.50	5.03	4.18	4.67	0.11
Economic growth share of infrastructure industries [(A/(A + B)) × 100]	0.7	3.6	14.1	11.8	−15.7

Source: Authors' calculations based on BEA's exact formula for component contributions to chained-dollar aggregate growth.

Note
a See the last row of Table 9.4.

IT Capital Stock and IT Expenditures

During the 1996–2000 growth spurt, the net stocks[9] of IT equipment and software for all industries grew by an average seven percent per year, from US$843 billion to US$1.2 trillion (see Table 9.7). During the period, about one-third of the total current cost of IT equipment and software assets in the US was owned by infrastructure industries. Particularly strong growth in net stocks of IT equipment occurred in the Transportation services industry (16 percent per year), the Airline industry (15 percent), and the Communications industry (12 percent).

In addition to investing dollars in IT capital assets and software, infrastructure industries spent significant sums on communications and computer services (see Table 9.8). Business expenses such as these are considered a cost of production and are not included in the industry's value of capital assets. Between 1996 and 2001, spending by infrastructure industries for communications increased by 23 percent, from US$114 billion in 1996 to US$164 billion in 2001. The majority (90 percent) of this spending on communications services derives from the communications industry itself. This means that firms/establishments in this industry purchase services from other firms/establishments within the same industry.

Table 9.6 Contribution of Infrastructure Industries to Price Growth

	1997	1998	1999	2000	2001
	Percentage points				
Railroads	0.003	0.015	−0.009	−0.010	0.010
Local and interurban transportation	0.002	0.012	−0.003	0.000	0.008
Trucking	0.113	0.106	0.018	−0.020	0.078
Water transportation	−0.001	0.006	0.012	0.000	0.008
Air transportation	0.041	0.064	−0.002	−0.027	−0.049
Pipelines, exc. Natural gas	−0.005	0.001	−0.001	0.003	0.006
Transportation services	0.009	−0.006	−0.001	0.016	0.001
Telephone and telegraph	0.015	0.007	−0.099	−0.150	−0.142
Radio and TV broadcasting	0.053	0.054	0.034	0.053	−0.066
Electric, gas, and sanitary services	0.050	0.089	−0.105	0.004	0.266
Price growth contributions of infrastructure industries (A)	0.25	0.33	−0.15	−0.13	0.12
Price growth contribution of the rest-of-US economy (B)	1.70	0.90	1.59	2.23	2.24
Percent					
Price growth = Growth of GDI implicit price deflator (A + B)[a]	1.95	1.23	1.44	2.10	2.36
Price growth share of infrastructure industries [(A/(A + B)) × 100]	12.8	27.1	−10.8	−6.3	5.1

Source: Authors' calculations based on BEA's exact formula for component contributions to chained-dollar aggregate growth.

Note

a The GDI implicit price deflator is the ratio of the total nominal GDI to the aggregate real GDI in Table 9.4. The price growth above is the same each year as in Table 9.3.

The infrastructure industry spending on computer services more than doubled over the time period. In 1998, infrastructure industries spent US$28 billion on computer services. In 2001, they spent US$64 billion. The spending estimates provided here do not include the "in-house" spending on communications or computer services, i.e. employees hired to perform communications and computer services. Total spending by infrastructure industries on computer services would be significantly higher if the in-house expenses were added.

IT Intensity

Infrastructure Industries Rank High

As we have just demonstrated, infrastructure industries invested heavily in IT equipment and software and might be considered the leading industries in terms of these investments. Our assessment, however, ranks industries in terms of their

Table 9.7 *Net Stocks of Information Technology Equipment in US Nonfarm Business Sector*

	SIC[a]	*1996*	*1997*	*1998*	*1999*	*2000*	*2001*
	Current cost (million dollars)						
Metal mining	10	1,472	1,407	1,320	1,172	1,094	1,024
Coal mining	12	720	764	747	762	791	825
Oil and gas extraction	13	11,667	12,345	12,688	12,658	12,845	13,026
Nonmetallic minerals, except fuels	14	624	684	740	811	889	952
Construction	15,16,17	5,510	5,987	6,456	7,267	7,743	7,683
Lumber and wood products	24	1,427	1,436	1,454	1,514	1,546	1,536
Furniture and fixtures	25	716	776	833	913	962	961
Stone, clay, and glass products	32	2,042	2,271	2,453	2,650	2,844	2,906
Primary metal industries	33	3,668	3,654	3,623	3,540	3,462	3,334
Fabricated metal products	34	4,160	4,377	4,650	4,857	5,053	5,054
Industrial machinery and equipment	35	14,534	16,479	18,033	19,351	20,388	20,462
Electronic and other electric equipment	36	19,281	19,752	19,626	19,765	21,255	21,659
Motor vehicles and equipment	371	4,908	5,268	5,636	5,782	5,779	5,690
Other transportation equipment	37exc371	5,561	6,008	6,601	6,927	6,809	6,548
Instruments and related products	38	12,303	13,068	13,959	14,843	15,694	15,919
Miscellaneous manufacturing industries	39	1,029	1,039	1,063	1,110	1,140	1,125
Food and kindred products	20	8,955	9,337	9,637	10,101	10,338	10,356
Tobacco products	21	899	993	892	827	750	686
Textile mill products	22	1,485	1,540	1,527	1,500	1,391	1,280
Apparel and other textile products	23	978	975	978	1,019	1,043	1,024
Paper and allied products	26	6,218	6,422	6,563	6,420	6,442	6,381
Printing and publishing	27	11,125	12,477	12,955	14,297	14,960	14,906
Chemicals and allied products	28	34,955	36,130	37,480	38,060	38,241	38,169
Petroleum and coal products	29	9,623	9,308	9,003	8,635	8,598	8,561
Rubber and miscellaneous plastics products	30	2,294	2,486	2,687	2,943	3,059	3,067
Leather and leather products	31	138	160	168	180	189	191
Railroad transportation	40	3,360	3,481	3,563	3,672	3,415	3,086
Local and interurban passenger transit	41	1,028	994	977	1,029	1,080	1,090
Trucking and warehousing	42	5,205	5,654	5,597	5,617	5,917	5,801
Water transportation	44	637	683	676	662	625	590
Transportation by air	45	13,917	17,109	19,453	21,216	24,624	27,432
Pipelines, except natural gas	46	1,839	2,037	1,890	1,927	2,155	2,252
Transportation services	47	8,690	9,887	11,715	13,662	17,050	18,007
Telephone and telegraph	481,482,489	180,817	191,390	192,825	205,360	236,968	243,741

Table 9.7 Continued

	SIC[a]	*1996*	*1997*	*1998*	*1999*	*2000*	*2001*
Radio and television	483,484	34,813	39,861	44,472	51,251	59,769	62,256
Electric, gas, and sanitary services	49	51,438	50,239	48,216	48,650	50,907	52,128
Wholesale trade	50,51	86,974	95,185	101,126	113,355	126,835	128,609
Retail trade	52–59	29,555	30,904	32,968	35,679	39,957	40,778
Depository institutions	60	36,594	39,171	42,799	47,137	49,490	48,359
Nondepository institutions	61	56,195	64,008	72,542	84,981	93,438	93,175
Security and commodity brokers	62	10,770	11,819	14,010	16,669	20,692	20,157
Insurance carriers	63	25,900	27,036	29,233	32,524	35,371	35,138
Insurance agents, brokers, and service	64	2,683	2,920	3,447	3,799	4,188	4,212
Holding and other investment offices	67	3,620	4,120	4,780	6,620	7,398	7,534
Hotels and other lodging places	70	2,270	2,384	2,395	2,671	2,755	2,726
Personal services	72	4,577	3,938	3,513	3,351	3,352	3,217
Business services	73	40,438	46,974	56,391	72,714	87,010	91,410
Auto repair, services, and parking	75	2,108	2,156	2,254	2,470	2,765	2,815
Miscellaneous repair services	76	2,078	2,430	2,369	2,539	2,908	3,046
Motion pictures	78	8,856	10,124	10,773	10,497	9,913	9,164
Amusement and recreation services	79	1,229	1,213	1,264	1,483	1,802	1,952
Health services	80	34,117	36,060	39,349	42,684	46,088	49,706
Legal services	81	4,284	4,777	5,212	5,780	6,244	6,213
Educational services	82	974	1,055	1,130	1,212	1,368	1,400
Other services	83,84,86,87,89	21,920	24,703	26,608	31,684	36,264	38,598
Sum		843,178	907,455	963,319	1,058,799	1,173,653	1,197,917
Sum of infrastructure industries' assets		301,744	321,335	329,384	353,046	402,510	416,383
Infrastructure share of IT assets		35.8%	35.4%	34.2%	33.3%	34.3%	34.8%

Source: Authors' calculations based on BEA data.

Note
a 1987 Standard Industrial Classification.

level of investments relative to their employment base. The intensity of an industry, in this analysis, is based on the annual level of the industry's current cost of IT equipment and software in relation to the number of employees within that industry. We define IT intensity as the ratio of the average value of the current cost of IT equipment and software owned by an industry in the 1996 to 2001 time period divided by the average employment in that industry divided by the value of IT equipment and software to employment of all industries.

Table 9.8 Communications and Computer Services Expenses of Infrastructure Industries

	1996	1997	1998	1999	2000	2001
	Communications ($ billions)					
Railroads	0.1	0.1	0.1	0.1	0.1	0.1
Trucking, highway passenger traffic	5.4	6.0	5.6	6.9	7.1	7.3
Water transportation	0.1	0.1	0.1	0.1	0.1	0.1
Air transportation	2.9	3.1	3.3	3.5	3.8	3.6
Pipelines	0.0	0.0	0.0	0.0	0.0	0.0
Transportation services	1.4	1.6	1.7	1.8	1.9	2.0
Communications	101.1	109.3	121.7	132.4	142.6	146.9
Electric utilities	1.1	1.1	1.1	1.4	1.5	1.5
Gas utilities	0.1	0.1	0.1	0.1	0.1	0.1
Water and sanitary services	1.6	1.6	1.7	1.8	2.1	2.2
Total	113.8	123.0	135.4	148.1	159.3	163.8
	Computer Services ($ billions)					
Railroads	0.9	1.1	1.3	1.4	1.6	1.6
Trucking, highway passenger traffic	2.1	2.7	3.2	4.0	4.8	4.8
Water transportation	0.3	0.3	0.4	0.5	0.6	0.6
Air transportation	5.2	6.6	8.0	9.3	11.3	11.0
Pipelines	0.1	0.1	0.1	0.1	0.2	0.2
Transportation services	4.0	5.1	6.4	7.7	8.7	9.4
Communications	8.9	11.2	14.5	17.7	21.0	22.7
Electric utilities	6.3	7.3	8.5	9.8	11.5	12.3
Gas utilities	0.3	0.5	0.4	0.5	0.8	0.8
Water and sanitary services	0.1	0.1	0.1	0.1	0.2	0.2
Total	28.2	35.0	42.9	51.1	60.7	63.6

Source: Authors' calculations based on input-output economic techniques.

Based on these criteria, five of the top ten IT intensive industries were infrastructure industries (see Table 9.9). Communications, both an infrastructure and an IT producing industry, ranked number 1 in terms of its "IT intensity" using this criterion. Other infrastructure industries with high rankings in its IT intensiveness were Pipelines (2); Radio and TV Broadcasting (4); Electric, gas and sanitary services (5); and Transportation services (9).

The remainder of the infrastructure industries ranked further down: Air transportation (ranked 17), Railroads (19), Water transportation (37), Trucking and warehousing (39), and Local and interurban passenger transit (45).

High IT Intensity Not Reflected in Infrastructure Industry Labor Productivity Growth Rates

Despite the heavy investment in IT equipment and software, in general, the productivity growth of the infrastructure industries was moderate and did not contribute substantially to overall labor productivity growth (see Table 9.10). As a group, these industries contributed only an average 0.2 percentage points to the

Table 9.9 *IT Intensity Ranking of US Nonfarm Business Industries*

	Rank	*1996–2001*		*1996*	*1997*	*1998*	*1999*	*2000*	*2001*
		Average intensity	*SIC*	*IT intensity*					
Telephone and telegraph	**1**	20.37	481,482,489	22.98	21.78	20.45	19.24	19.04	18.71
Pipelines, except natural gas	**2**	14.32	46	14.25	15.12	14.64	13.93	14.40	13.60
Nondepository institutions	**3**	12.08	61	12.39	12.16	11.64	11.93	12.58	11.77
Radio and television	**4**	11.47	483,484	10.07	10.93	11.46	11.95	12.25	12.18
Petroleum and coal products	**5**	6.61	29	7.57	7.16	6.82	6.24	5.98	5.88
Electric, gas, and sanitary services	**6**	5.67	49	6.40	6.08	5.73	5.38	5.25	5.20
Oil and gas extraction	**7**	3.81	13	4.02	3.85	3.84	4.10	3.69	3.33
Chemicals and allied products	**8**	3.51	28	3.72	3.68	3.70	3.50	3.24	3.21
Transportation services	**9**	2.88	47	2.38	2.47	2.73	2.94	3.31	3.46
Metal mining	**10**	2.62	10	2.96	2.71	2.71	2.50	2.38	2.47
Holding and other investment offices	**11**	2.25	67	1.65	1.82	2.03	2.64	2.66	2.68
Security and commodity brokers	**12**	2.21	62	2.10	2.04	2.18	2.28	2.40	2.22
Tobacco products	**13**	2.19	21	2.44	2.58	2.30	2.16	1.92	1.76
Depository institutions	**14**	2.19	60	2.07	2.11	2.23	2.30	2.26	2.15
Motion pictures	**15**	2.07	78	2.28	2.35	2.32	2.03	1.78	1.64
Insurance carriers	**16**	1.98	63	1.94	1.92	1.96	2.00	2.05	2.00
Transportation by air	**17**	1.72	45	1.45	1.67	1.75	1.72	1.77	1.94
Instruments and related products	**18**	1.63	38	1.59	1.60	1.64	1.67	1.64	1.63
Railroad transportation	**19**	1.61	40	1.73	1.74	1.70	1.65	1.48	1.36
Wholesale trade	**20**	1.57	50,51	1.50	1.53	1.54	1.58	1.62	1.66
Electronic and other electric equipment	**21**	1.17	36	1.27	1.23	1.17	1.13	1.09	1.14
Paper and allied products	**22**	0.94	26	1.00	0.99	0.99	0.91	0.86	0.87
Printing and publishing	**23**	0.88	27	0.84	0.88	0.88	0.92	0.89	0.89
Coal mining	**24**	0.85	12	0.81	0.84	0.83	0.85	0.89	0.87
Industrial machinery and equipment	**25**	0.83	35	0.76	0.80	0.83	0.86	0.85	0.88
Business services	**26**	0.74	73	0.64	0.65	0.70	0.78	0.80	0.86
Other transportation equipment	**27**	0.73	37exc371	0.74	0.74	0.75	0.75	0.71	0.69
Nonmetallic minerals, except fuels	**28**	0.69	14	0.64	0.66	0.68	0.69	0.70	0.73
Miscellaneous repair services	**29**	0.67	76	0.63	0.70	0.65	0.65	0.71	0.71
Food and kindred products	**30**	0.57	20	0.59	0.59	0.59	0.57	0.54	0.53
Motor vehicles and equipment	**31**	0.54	371	0.56	0.56	0.57	0.53	0.50	0.51
Legal services	**32**	0.52	81	0.49	0.52	0.53	0.53	0.52	0.50
Primary metal industries	**33**	0.50	33	0.57	0.54	0.51	0.48	0.43	0.44
Insurance agents, brokers, and service	**34**	0.46	64	0.41	0.42	0.47	0.48	0.49	0.47
Stone, clay, and glass products	**35**	0.43	32	0.42	0.43	0.44	0.44	0.43	0.44
Health services	**36**	0.43	80	0.42	0.42	0.43	0.43	0.43	0.43
Water transportation	**37**	0.36	44	0.42	0.42	0.39	0.35	0.30	0.28
Other services, n.e.c.	**38**	0.36	83,84,86, 87,89	0.34	0.35	0.35	0.37	0.37	0.38
Trucking and warehousing	**39**	0.32	42	0.36	0.37	0.34	0.31	0.29	0.28
Fabricated metal products	**40**	0.30	34	0.32	0.31	0.31	0.30	0.29	0.30
Personal services	**41**	0.30	72	0.43	0.35	0.30	0.26	0.24	0.22
Miscellaneous manufacturing industries	**42**	0.27	39	0.29	0.28	0.28	0.27	0.25	0.26
Rubber and miscellaneous plastics products	**43**	0.27	30	0.26	0.26	0.27	0.28	0.26	0.27
Textile mill products	**44**	0.25	22	0.26	0.26	0.26	0.26	0.23	0.23
Local and interurban passenger transit	**45**	0.23	41	0.28	0.25	0.23	0.22	0.21	0.20
Leather and leather products	**46**	0.21	31	0.16	0.19	0.20	0.22	0.23	0.26
Auto repair, services, and parking	**47**	0.19	75	0.20	0.19	0.19	0.18	0.18	0.18
Lumber and wood products	**48**	0.18	24	0.20	0.19	0.18	0.17	0.16	0.17
Retail trade	**49**	0.17	52–59	0.17	0.17	0.17	0.17	0.17	0.17
Furniture and fixtures	**50**	0.16	25	0.16	0.16	0.16	0.16	0.15	0.16
Hotels and other lodging places	**51**	0.15	70	0.16	0.15	0.15	0.15	0.14	0.13
Apparel and other textile products	**52**	0.14	23	0.13	0.13	0.13	0.14	0.15	0.16
Construction	**53**	0.10	15,16,17	0.11	0.11	0.11	0.11	0.10	0.10
Amusement and recreation services	**54**	0.10	79	0.10	0.09	0.09	0.10	0.10	0.11
Educational services	**55**	0.06	82	0.06	0.06	0.06	0.06	0.06	0.05

Source: Authors' calculations based on BEA data on IT equipment (ITEQ) and full-time equivalent (FTE) workers.

Note
In this chapter, IT intensity is the *ratio* of ITEQ to FTE for each industry divided by the *overall ratio* of ITEQ to FTE for all industries. The infrastructure industries are shown in bold.

Table 9.10 *Labor Productivity Growth in US Nonfarm Business Industries*

	1996–2001	*SIC*	*1997*	*1998*	*1999*	*2000*	*2001*	*1997–2001*
	Average rank				*Percentage points*			*Average*
Telephone and telegraph	**20.37**	**481,482,489**	**0.004**	**0.171**	**0.288**	**0.320**	**0.311**	**0.219**
Pipelines, except natural gas	**14.32**	**46**	**0.008**	**0.003**	**0.003**	**−0.002**	**−0.006**	**0.001**
Nondepository institutions	12.08	61	0.174	−0.029	0.066	0.189	0.136	0.107
Radio and television	**11.47**	**483,484**	**−0.019**	**−0.005**	**0.009**	**0.014**	**0.105**	**0.021**
Petroleum and coal products	6.61	29	−0.082	0.018	0.140	−0.074	−0.103	−0.020
Electric, gas, and sanitary services	**5.67**	**49**	**−0.092**	**−0.122**	**0.228**	**0.077**	**−0.287**	**−0.039**
Oil and gas extraction	3.81	13	0.032	0.004	−0.052	−0.205	0.020	−0.040
Chemicals and allied products	3.51	28	0.179	−0.147	0.081	0.031	−0.051	0.019
Transportation services	**2.88**	**47**	**−0.010**	**0.005**	**0.025**	**0.025**	**−0.009**	**0.008**
Metal mining	2.62	10	0.007	0.023	0.018	−0.005	0.018	0.012
Holding and other investment offices	2.25	67	0.027	0.211	−0.126	0.089	0.245	0.089
Security and commodity brokers	2.21	62	0.284	0.451	0.098	0.354	0.249	0.287
Tobacco products	2.19	21	−0.016	−0.038	−0.158	−0.027	−0.060	−0.060
Depository institutions	2.19	60	0.084	0.183	0.329	0.233	0.050	0.176
Motion pictures	2.07	78	−0.009	0.019	−0.014	−0.009	0.039	0.005
Insurance carriers	1.98	63	0.198	−0.048	−0.059	0.244	−0.069	0.053
Transportation by air	**1.72**	**45**	**0.055**	**−0.029**	**0.022**	**0.016**	**−0.082**	**−0.004**
Instruments and related products	1.63	38	−0.079	−0.021	−0.002	0.063	−0.082	−0.024
Railroad transportation	**1.61**	**40**	**−0.008**	**−0.003**	**−0.005**	**0.045**	**0.009**	**0.008**
Wholesale trade	1.57	50,51	0.707	1.006	0.506	0.439	0.243	0.580
Electronic and other electric equipment	1.17	36	0.432	0.359	0.439	0.457	0.242	0.386
Paper and allied products	0.94	26	0.041	−0.028	0.009	−0.032	−0.025	−0.007
Printing and publishing	0.88	27	−0.054	−0.026	0.093	−0.002	−0.107	−0.019
Coal mining	0.85	12	0.002	0.015	0.027	−0.011	0.018	0.010
Industrial machinery and equipment	0.83	35	0.286	0.463	0.198	0.443	−0.137	0.251
Business services	0.74	73	0.021	−0.224	0.004	−0.437	0.370	−0.053
Other transportation equipment	0.73	37exc371	0.021	−0.015	0.104	0.019	0.048	0.035
Nonmetallic minerals, except fuels	0.69	14	0.006	0.006	−0.003	0.021	0.003	0.006
Miscellaneous repair services	0.67	76	−0.015	0.001	−0.027	0.024	−0.065	−0.016
Food and kindred products	0.57	20	−0.007	−0.098	0.129	−0.150	−0.069	−0.039
Motor vehicles and equipment	0.54	371	0.065	0.213	0.023	0.035	−0.043	0.059
Legal services	0.52	81	0.089	0.022	0.035	0.071	0.032	0.050
Primary metal industries	0.50	33	0.028	0.023	0.053	−0.013	0.023	0.023
Insurance agents, brokers, and service	0.46	64	−0.017	0.032	0.042	−0.044	0.011	0.005
Stone, clay, and glass products	0.43	32	0.046	−0.015	0.008	−0.012	−0.037	−0.002
Health services	0.43	80	−0.217	−0.119	0.026	0.138	−0.079	−0.050
Water transportation	**0.36**	**44**	**0.013**	**−0.004**	**−0.022**	**0.017**	**0.000**	**0.001**
Other services, n.e.c.	0.36	83,84,86, 87,89	−0.124	−0.113	−0.093	−0.131	−0.178	−0.128
Trucking and warehousing	**0.32**	**42**	**−0.077**	**0.025**	**0.017**	**0.020**	**−0.059**	**−0.015**
Fabricated metal products	0.30	34	0.014	−0.028	0.003	0.061	−0.068	−0.003
Personal services	0.30	72	0.030	0.061	−0.032	−0.034	−0.021	0.001
Miscellaneous manufacturing industries	0.27	39	0.020	−0.002	0.036	0.057	−0.019	0.019
Rubber and miscellaneous plastics products	0.27	30	0.038	0.023	0.033	0.043	0.000	0.028
Textile mill products	0.25	22	0.007	0.007	0.021	0.026	0.025	0.018
Local and interurban passenger transit	**0.23**	**41**	**0.005**	**−0.001**	**0.006**	**−0.004**	**−0.005**	**0.000**
Leather and leather products	0.21	31	0.007	−0.001	−0.002	0.011	0.003	0.004
Auto repair, services, and parking	0.19	75	−0.026	0.057	0.050	0.003	0.013	0.019

Table 9.10 Continued

	1996–2001	*SIC*	*1997*	*1998*	*1999*	*2000*	*2001*	*1997–2001*
Lumber and wood products	0.18	24	−0.019	−0.014	−0.015	0.030	0.005	−0.003
Retail trade	0.17	52–59	0.573	0.546	0.112	0.286	0.496	0.403
Furniture and fixtures	0.16	25	0.019	−0.008	−0.004	−0.001	0.009	0.003
Hotels and other lodging places	0.15	70	−0.071	−0.067	−0.046	0.021	−0.024	−0.037
Apparel and other textile products	0.14	23	0.036	0.042	0.033	0.082	0.043	0.047
Construction	0.10	15,16,17	−0.188	0.077	−0.084	−0.135	−0.146	−0.095
Amusement and recreation services	0.10	79	0.008	0.001	−0.080	−0.080	0.022	−0.026
Educational services	0.06	82	−0.049	−0.052	−0.051	−0.057	−0.047	−0.051
Sum of labor productivity contributions of all nonfarm business industries			**2.386**	**2.808**	**2.441**	**2.538**	**0.910**	**2.217**
Sum of labor productivity contributions of all infrastructure industries			**−0.120**	**0.039**	**0.571**	**0.528**	**−0.023**	**0.199**
Infrastructure share of productivity growth			**−5.0%**	**1.4%**	**23.4%**	**20.8%**	**−2.5%**	**9.0%**

Source: Authors' calculations based on BEA data on gross domestic product (GDP) and full-time equivalent (FTE) worker by industry. Labor productivity is defined as the ratio of GDP to FTE. The calculations cover only the above 55 industries that roughly correspond to the US nonfarm business sector, excluding real estate. The infrastructure industries are shown in bold.

Note
The decomposition of overall labor productivity growth into individual industry contributions follows the same methodology used in the Economics and Statistics Administration, US Department of Commerce, *Digital Economy 2002*, chapter 4, available at http://www.esa.doc.gov/pdf/DE2002r1.pdf.

2.2 percent productivity growth rate in the 1997 to 2001 period. That means that the remaining industries in the economy contributed over 1.9 percentage points. By contrast, in 1999 and 2000, though, infrastructure industries contributed over 20 percent of the overall labor productivity growth rate.

Communications consistently out-performed the other infrastructure industries. Communications contributed an average 0.2 percentage points out of the average 2.2 labor productivity growth, a ten percent share, during the 1997 to 2001 time period. There were only five other industries, none of which were infrastructure industries, which did better than Communications in terms of their labor productivity growth contribution. They were Industrial machinery and equipment (11 percent), Security and commodity brokers (13 percent), Electronic components (17 percent), Retail trade (18 percent), and Wholesale trade 26 percent). All other infrastructure industries were either mildly positive or mildly negative, on the average, over the period.

IT Technologies Associated with Displacement of Jobs in Infrastructure Industries: Effects on Labor Productivity

We determined that in the infrastructure sector, the relative number of workers with certain IT sensitive occupations declined while the infrastructure industry, in general, increased. We were unable, however, to establish causality between IT technologies and displacement of workers. Despite the lack of proof of causality, our results provide some interesting speculation that the reduced demand for these workers had positive effects on the labor productivity of infrastructure industries. In other words, labor productivity in infrastructure industries would have been less except for this reduced demand for workers whose occupations may have been displaced because of IT.

Between 1997 and 2000,[10] a period of rapid growth and low unemployment, IT substituted for some job opportunities in IT-affected job categories and made remaining workers more productive. Based on an analysis of employment by industry by occupation, we estimate that between 1997 and 2000, the growth in the use of IT technologies is associated with displaced workers in 43 occupations (see Figure 9.3).[11] These include not only clerks and secretaries, but also purchasing managers and several types of operators. They fall in administrative, managerial, and production categories.

In our analysis, we looked at the number of workers in occupations we consider to be considerably affected by IT technologies. We've termed them to be IT-displaceable jobs. We then looked at the number of workers in 2000 in these occupations. Since we know the output of these industries, we are able to determine the number of workers in these occupations per unit of output for 1997 and

Purchasing managers	Buyers and purchasing agents
Credit analysts	Insurance claims adjusters
Insurance underwriters	Loan counselors
Insurance sales agents	Securities, commodities/financial services sales
Adjustment clerks	Bill collectors
Insurance claims clerks	Telephone operators
Other communications operators	Computer operators
Peripheral computer equip. oper.	New accounts clerks, banking
Mail clerks, except mail machine Operators	Procurement clerks
	Production and planning clerks
Shipping and receiving clerks	All other recording clerks
Advertising clerks	Brokerage clerks
Correspondence clerks	File clerks
Financial records clerks	Library assistants
Human resource assistants	Statement clerks
Court reporters	Secretaries
Word processors and typists	Bank tellers
Court clerks	Credit and loan authorizers
Data entry keyers	Office machine operators
Office clerks	Proofreaders
Statistical clerks	All other clerks
Printing workers, precision	Printing, binding, and related workers

9.3 *Occupations Displaced by IT Technologies*
Source: Authors' designation of occupations based on Bureau of Labor Statistics forecasts of IT occupation demand. See endnote 11.

2000. We then calculated the ratio of IT displaceable workers in 1997 and 2000. We attributed the difference to IT technologies displacing jobs. We chose the 1997 to 2000 period as our period of analysis since there was rapid growth in output in infrastructure industries and layoffs would not be attributed to declines in revenues.

Jobs in these 43 occupations, over the period, increased at a 1.2 percent average annual rate. In contrast, total jobs – jobs in all occupations – increased at a 2.4 percent annual rate.[12] In 1997, there were 18.1 million workers in these occupations across all industries. This amounts to about 13.2 percent of the total 135 million jobs in the US By 2000, there were 18.8 million workers or 12.8 percent out of 145 million jobs. Despite the relatively small 0.4 percentage point difference (13.2 percent minus 12.8 percent), the result is that 700,000 workers with these occupations were not needed in 2000 as they were in 1997.

Table 9.11 provides industry level estimates of IT's influence on employment in infrastructure industries and how the displacement of jobs in these occupational categories has affected overall industry productivity. For all industries,

Table 9.11 *Jobs in Infrastructure Industries Affected by IT Technology, 1997–2000*

Industry	*IT-affected Jobs 1997 (×000)*	*IT-affected Share of Employment 1997 (percent)*	*Estimated Number of IT-affected Jobs Saved in 2000 (×000)*	*Labor productivity Growth, 1997–2000*[b]	
				Actual Growth (percent)	*Without IT-affected Jobs Saved (percent)*
Electric utilities	66.4	14.7	−3.1	−1.8	−1.5
Gas utilities	27.1	15.5	0.9	0.1	−0.1
Water and sewage	25.7	10.4	−1.3	−2.3	−2.2
Air transportation	100.6	8.8	3.1	2.0	1.9
Rail transportation	13.3	5.9	1.3	4.0	3.8
Water transportation	13.1	7.0	0.9	1.8	1.7
Trucking and interurban ground transportation	202.9	8.2	7.9	0.6	0.6
Transportation services	84.5	17.6	1.1	0.1	0.0
Communications	377.0	26.0	56.4	3.2	2.0
Average all industries, including infrastructure				2.5	1.6

Source: Bureau of Labor Statistics data and authors calculations of jobs saved and labor productivity.

Notes
a Number of additional jobs in 2000 needed by industries if employment by IT-affected job share of real output kept at the 1997 level.
b Labor productivity measured in real GDP by industry and growth in compound annual rates.

including infrastructure industries, we estimate that the labor productivity rate would have been 0.9 percentage points less if IT technology had not contributed to labor productivity. Put another way, if IT technologies had not provided greater labor efficiencies, infrastructure industries would have needed to employ workers in these occupations as they had in the beginning of the growth cycle.

In 1997, 26 percent of the jobs in the Communications industry (about 377 thousand jobs) were in the IT-displaceable job categories, i.e. those listed in Figure 9.3. In 1997, there were 1.5 million workers in the Communications industry. By 2000, there were 1.8 million workers in the Communications industry, but only 23 percent were in the 43 IT occupations – about 423,000 workers. Since the output of the industry continued to rise (Table 9.4) between 1997 and 2000, we would expect that without the improvement in labor efficiencies in these occupations that the Communications industry would have required an additional 52,000 workers, and, thus, maintain the 26 percent of the workforce that it had in 1997.

Between 1997 and 2000, the Communications industry's labor productivity increased by an average 3.2 percent per year. If the industry had required the same number of workers in these occupations per unit of output as it had in 1997, then its labor productivity rate would have been two percent – a 1.2 percentage point reduction in the productivity rate.

For the remainder of the infrastructure industries, the displacement of workers in these job categories had very little, if not negative, influence on their labor productivity growth in the time period. This means that these industries tended to hold on to or add workers in these occupations over the period.

Conclusion

The infrastructure sector has indeed become digital. The data support the contention that an underlying digital infrastructure for the infrastructure sector exists and it consists of that industry's investments in IT equipment and software and its consumption of IT services. The most significant finding in all the quantitative economic analyses done for this chapter is that infrastructure industries which account for eight to nine percent of the nation's GDP own "one-third" of the nation's capital stock in IT equipment and software.

More work needs to be done to completely understand the phenomenon of the commitment by the infrastructure industries to invest so heavily in IT equipment and software when the rewards such as improvement in labor productivity are not reflected in the industry data. For certain, IT investments alone by the infrastructure industries have not been sufficient to raise productivity in them.

It may be perhaps that the levels of employment in these industries are not as elastic as those in other parts of manufacturing and services. It may be that rising prices in these sectors along with an inelastic demand, particularly in utilities and communications, makes the dynamics of supply and demand act differently. In any event, the data presented in this chapter has shown that the nation's infrastructure industries have truly become digitized.

Notes

1 The analysis and conclusions in this chapter are those of the authors and do not in any way represent the views of the US Department of Commerce.
2 The average labor productivity growth rate for infrastructure industries between 1993 and 1996 was equal to its productivity growth rate between 1997 and 2000. Thus, no labor productivity improvements were seen as a result of the heavy investment in IT.
3 See *Digital Economy 2002*, Appendices, Appendix to Chapter III, p. 1, "Defining Information Technology Producing Industries," available from www.esa.doc.gov
4 See *Digital Economy 2002*, p. 31, Chapter IV, Industry-Level Effects of Information Technology Use on Productivity and Inflation, US Department of Commerce, Economics and Statistics Administration, February 2002.
5 Bureau of Economic Analysis, US Department of Commerce, Table 5.3.5, Private Fixed Investment by Type (A) (Q) and Table 5.3.6, Real Private Fixed Investment by Type, Chained Dollars (A) (Q), www.bea.doc.gov.
6 See "Gross Domestic Product by Industry for 1999–2001" by Robert J. McCahill and Brian C. Moyer, Bureau of Economic Analysis, Survey of Current Business, November 2002.
7 For BEA's use of hedonic price indexes, see "A Note on the Impact of Hedonics and Computers on Real GDP" by Steven Landefeld and Bruce Grim, Survey of Current Business, December 2000 and "Measuring the New Economy" by Steven Landefeld and Barbara Fraumeni, Survey of Current Business, March 2001.
8 See "The Impacts of ICT [Information and Communication Technology] on Economic Performance – An International Comparison at Three Levels of Analysis" by Dirk Pilat and Andrew Wyckoff, OECD, prepared for the conference "Transforming Enterprise," US Department of Commerce, January 2003; see also "ICT and Economic Growth," OECD, September 2003, www.oecd.org.
9 Net stocks of IT equipment and software are the current cost value of IT equipment and software plus new investments minus depreciation. Types of IT assets included in this analysis are computers (mainframe, personal, terminals tape drives, storage devices, and printers), direct access storage devices, integrated systems, software (prepackaged, custom, and own-account), communications equipment, instruments, photocopy equipment, office and accounting equipment.
10 The 1997 to 2000 period was used here because the BLS employment by occupation time series prior to 1997 was not consistent with the time series after 1997 and was a period of rapid IT-related economic growth.
11 Occupations affected by IT technology were identified based on descriptions of factors affecting employment in the BLS' Occupation Outlook Handbook. BLS compiles occupation employment data for more than 800 occupations.
12 Based on BLS data and total jobs in industry, government enterprises (post office), Federal, military and civilian government, and State and local government.

References

Bureau of Economic Analysis, US Department of Commerce, Various Tables, Online. Available at: www.bea.doc.gov (accessed January 15, 2004).

Bureau of Labor Statistics (BLS), *Occupational Outlook Handbook*, Online. Available at: www.bls.gov/oco/ (accessed January 15, 2004).

Economics and Statistics Administration, US Department of Commerce (2002) *Digital Economy 2002*, Online. Available at: www.esa.doc.gov (accessed January 8, 2004).

Economics and Statistics Administration, US Department of Commerce (2003) *Digital Economy 2003*, Online. Available at: www.esa.doc.gov (accessed January 8, 2004).

Landefeld, J.S. and Parker, R.P. (1997) "BEA's Chain indexes, Time Series and Measures of Long-term Economic Growth," *Survey of Current Business*, Vol. 77, No. 5, pp. 58–68.

Landefeld, J.S. and Grim, B. (2000) "A Note on the Impact of Hedonics and Computers on Real GDP," in the US Department of Commerce, *Survey of Current Business*, December.

Landefeld, J.S. and Fraumeni, B. (2001) "Measuring the New Economy," in the US Department of Commerce, *Survey of Current Business*, March.

McCahill, R.J. and Moyer, B.C. (2002) "Gross Domestic Product by Industry for 1999–2001," Bureau of Economic Analysis, Survey of Current Business, November.

Organisation for Economic Cooperation and Development (2003) "ICT (Information and Communications Technology) and Economic Growth: Evidence from OECD Countries, Industries, and Firms," Online. Available at: www.oecd.org (accessed January 8, 2004).

Pilat, D. and Wyckoff, A. (2003) "The Impacts of ICT (Information and Communications Technology) on Economic Performance – An International Comparison at Three Levels of Analysis," Organization for Economic Cooperation and Development, prepared for the conference "Transforming Enterprise," US Department of Commerce, January.

CHAPTER TEN

Infrastructure and IT Dimensions in the Developing World

Carlos Restrepo

Introduction

Although most applications of information technology (IT) to infrastructure systems are being developed and deployed in developed countries, it is possible that in the near future the greatest need will be for their use in developing countries. Current population growth and urbanization rates indicate that in the Twenty-First Century most people will live in what are currently known as developing countries and that the majority of the world's large urban areas will also be located in developing countries. Many of these rapidly growing urban areas already face important deficits in the provision of infrastructure systems such as access to clean water and wastewater treatment, electricity and transportation and these defecits will increase in the coming years. This chapter presents data that demonstrate enormous disparities in access to infrastructure services around the world and examines how IT can help meet some of the infrastructure needs in developing countries. The infrastructure sectors considered include water and wastewater treatment, transportation, energy and electricity, and telecommunications and information technology. For the purposes of this chapter, developing countries are defined as low- and middle-income countries. The World Bank (2001: 271) defines these countries as having a Gross National Product (GNP) per capita of US$9,266 or less.

Demographic Trends and Income

Current demographic trends suggest that in the next few decades the greatest demand for infrastructure systems will be in developing countries, where most of the world's population will live. The United Nations (2003) estimates that between now and 2050 the population of the more developed regions of the world will only grow by about 0.25 percent per year, whereas the developing countries will experience a growth rate of 1.46 percent. The 49 poorest countries are expected to grow at a rate of 2.4 percent per year. By 2050 most developed countries will be experiencing negative population growth rates. In Japan and Italy the population in 2050 will be approximately 14 percent and 20 percent smaller than it is today according to these projections. And some countries in Eastern Europe will observe decreases between 30–50 percent (United Nations 2003).

These trends are also true for urban areas. According to estimates by the United Nations, the percentage of the world's population living in urban areas will rise from 47 percent in 2000 to over 60 percent by 2030. Moreover, the UN expects that practically all population growth in the world during this period will take place in urban areas and most of this population increase will take place in urban areas in developing countries. By the year 2015 only three of the top twenty most populous urban areas in the world will be in developed countries. These projections are shown in Table 10.1, which compares the population of the world's largest urban agglomerations in 1995 with estimates for 2015 (World

Table 10.1 *Population Growth and Magnitude of Population of the World's Largest Urban Areas, 1995–2015*

City	*Population in 1995*	*Projected population in 2015*	*Percent growth*
Lagos, Nigeria	10.29	24.61	139.2
Dhaka, Bangladesh	8.55	19.49	128
Karachi, Pakistan	9.73	19.38	99.2
Bombay, India	15.14	26.22	73.2
Delhi, India	9.95	16.86	69.4
Jakarta, Indonesia	8.62	13.92	61.5
Metro Manila, the Philippines	9.29	14.66	57.8
Cairo, Egypt	9.69	14.42	48.8
Calcutta, India	11.92	17.31	45.2
Tianjin, China	9.42	13.53	43.6
Beijing, China	11.3	15.57	37.8
Shanghai, China	13.58	17.97	32.3
Sao Paulo, Brazil	16.53	20.32	22.9
Buenos Aires, Argentina	11.8	13.86	17.5
Rio de Janeiro, Brazil	10.18	11.86	16.5
Mexico City, Mexico	16.56	19.18	15.8
Los Angeles, USA	12.41	14.22	14.6
Seoul, Korea, Rep.	11.61	12.98	11.8
New York, USA	16.33	17.6	7.8
Tokyo, Japan	26.96	28.89	7.2
Paris, France	9.52	9.69	1.8
Moscow, Russian Federation	9.27	9.3	0.3
Osaka, Japan	10.61	10.61	0

Source: Adapted from World Resources Institute *et al.* (1998), Table HW1, p. 147.

Resources Institute *et al.* 1998: 147). The cities are ranked by population growth rates. According to these figures, cities such as Lagos, Nigeria and Dhaka, Bangladesh will more than double their populations during this period. On the other hand, large cities in developed countries will undergo modest population changes, and some, such as Osaka, Moscow, and Paris will see little or no growth.

These differences in population trends between developed and developing countries are shown for a sample of countries from around the world in Figures 10.1 and 10.2. These figures show cities from countries with a diversity of income levels. Countries towards the left side of the x-axis represent poor countries and those toward the right side of the figures are some of the wealthiest countries in the world. Figure 10.1 clearly shows that urban growth rates in the poorer countries are much higher than in the wealthier countries for the three periods shown: 1980–1985, 2000–2005 and 2020–2025 (World Resources Institute *et al.* 1998: 274–5). Figure 10.2 shows the percentage of the population living in urban agglomerations with a population of 750,000 people or more. In most of the developing countries this percentage will increase between 1995 and 2015 whereas in the developed countries the percentage tends to decrease (World Resources Institute *et al.* 1998: 274–5). Given its geographical characteristics, Singapore is the only country in the sample with a 100 percent urban population.

In addition to these differences in demographic trends between developed and developing countries, income per capita figures around the world also continue to show striking differences. As Figure 10.3 shows, countries in the European Union, as well as Australia and the United States enjoy an annual gross

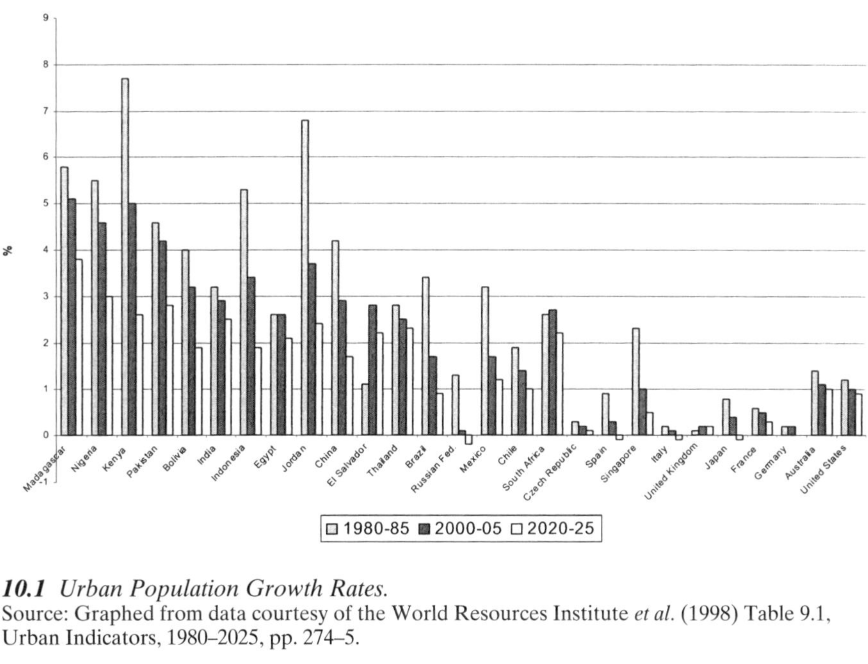

10.1 *Urban Population Growth Rates.*
Source: Graphed from data courtesy of the World Resources Institute *et al.* (1998) Table 9.1, Urban Indicators, 1980–2025, pp. 274–5.

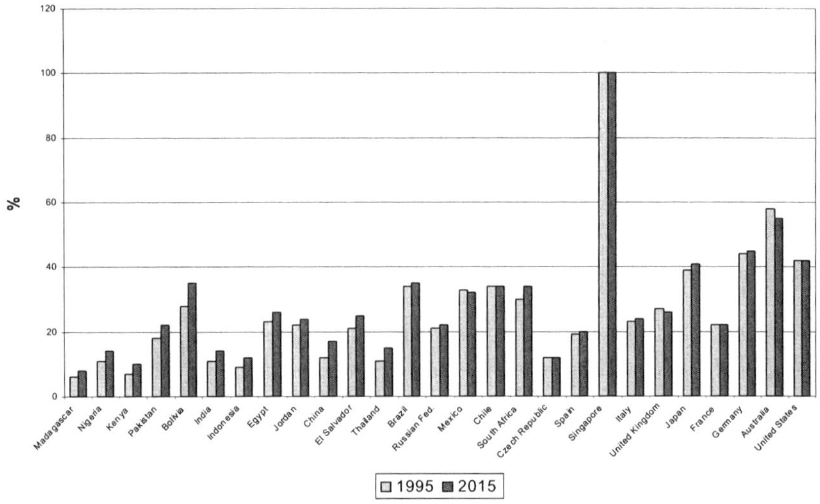

10.2 *Population in Urban Agglomerations Around the World.*
Source: Graphed from data courtesy of the World Resources Institute *et al.* (1998) Table 9.1, Urban Indicators, 1980–2025, pp. 274–5.

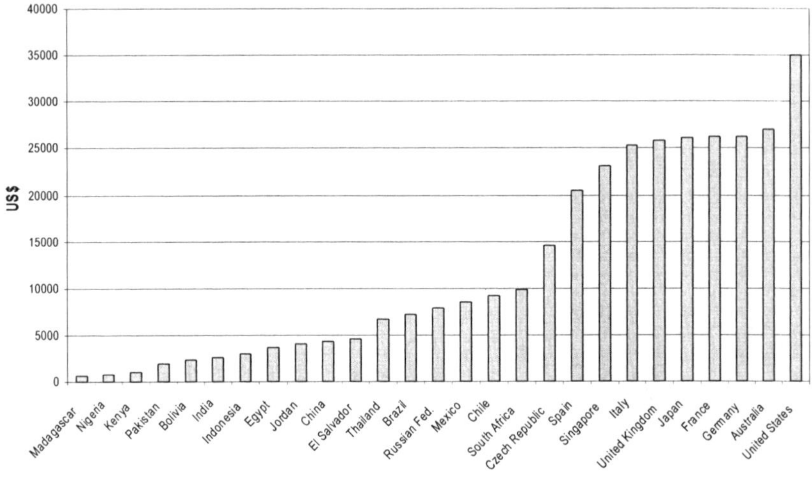

10.3 *Gross National Income Per Capita.*
Source: Graphed from data courtesy of the World Bank (2000a) Table 9, pp. 246–7.

national income per capita above US$25,000. At the other end of the scale, countries such as Madagascar and Nigeria have estimated gross national incomes per capita below US$1,000 (World Bank 2000a: 246–7). These figures show GNP per capita measured at purchasing power parity (PPP) exchange rates. The GNP per capita figures are converted to international dollars, which allows a standard comparison of real price levels between countries. Using PPP exchange rates means that one international dollar has the same purchasing power over a country's GNP that the US dollar has over US GNP (World Bank 2001).

Access to, and use of infrastructure services around the world also show large inequalities. Some trends over time for infrastructure use in developed countries are described in Chapter 1 of this volume. The following sections discuss differences in access to, and use of infrastructure services for a sample of countries similar to those in Figure 10.3. Indicators such as vehicle car ownership per 1,000 persons, percentage of paved roads, electricity consumption per capita, percentage of losses in electricity grids, telephone lines and mobile phones per 1,000 persons, Internet hosts per 10,000 persons, percentage of the population with access to safe water, and percentage of wastewater treated for selected cities around the world are used to describe these inequalities and opportunities are suggested where IT can be used to improve infrastructure service provision in developing countries and overcome some of the traditional problems of infrastructure service provision.

Water and Wastewater

Although developing countries made great strides in the last couple of decades to improve access to safe water, much work remains to be done. The most important difference between developed and developing countries in this respect is access to safe water in rural areas. Many developing countries have rates of access to safe water in excess of 80 percent for their urban populations. However, in rural areas the percentage tends to be much lower. Figure 10.4 shows these rates for 1995 for

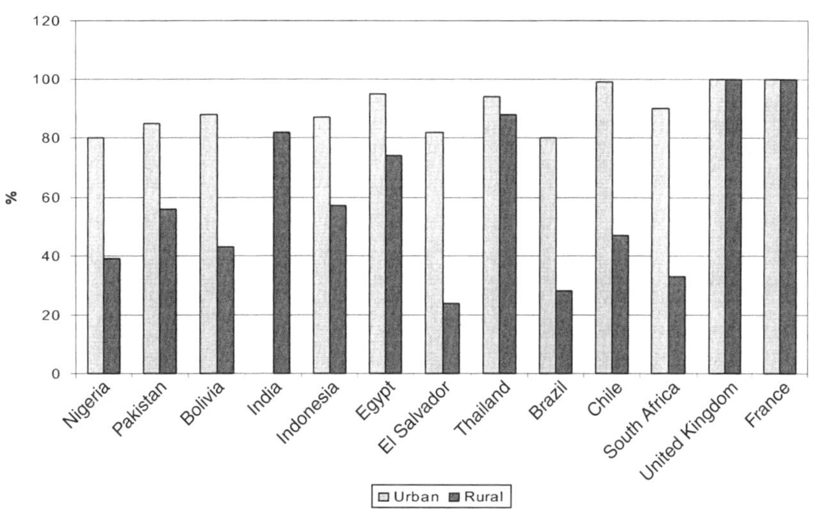

10.4 *Access to Safe Water, 1995.*
Source: Graphed from data courtesy of the World Bank (1999) Table 9, pp. 246–7.

a sample of countries (World Bank 1999: 246–7). Developed countries such as the United Kingdom and France report that 100 percent of their urban and rural populations have access to safe water. In other parts of the world that is not the case. In Chile, for example, almost 100 percent of the urban population had access to safe water in 1995, but the rate in rural areas only reached 47 percent. Similarly, other Latin American countries such as Bolivia, Brazil and El Salvador have much higher rates of access to safe water in urban areas compared to rural areas.

Another important difference between developed and many developing countries is the percentage of wastewater that is treated before being discharged into natural bodies of water. Figure 10.5 shows figures for a sample of cities from 1993. Some cities in developed countries such as Melbourne, New York City and Seattle report wastewater treatment rates of 100 percent. Moscow, Cairo and Nairobi also report very high rates of wastewater treatment. However, most of the cities from developing countries in the sample have much lower rates. Shanghai and Amman reported rates of 58 percent and 51 percent respectively in 1993. Rio de Janeiro treated about 23 percent of its wastewater in the same year. Other cities such as Antananarivo, Lahore, La Paz, San Salvador and Santiago treated very little or none of their wastewater in the early 1990s (World Resources Institute *et al.* 1998: 278–9). In these cities it will be important to increase coverage of this service in the upcoming decades and an important research question is how to include IT-based wastewater treatment technologies to meet their needs.

Another problem with water infrastructure that is often observed in developing countries is high rates of unaccounted for water losses due to leakages and illicit connections. Variation among cities in Asia is very high. For example, cities such as

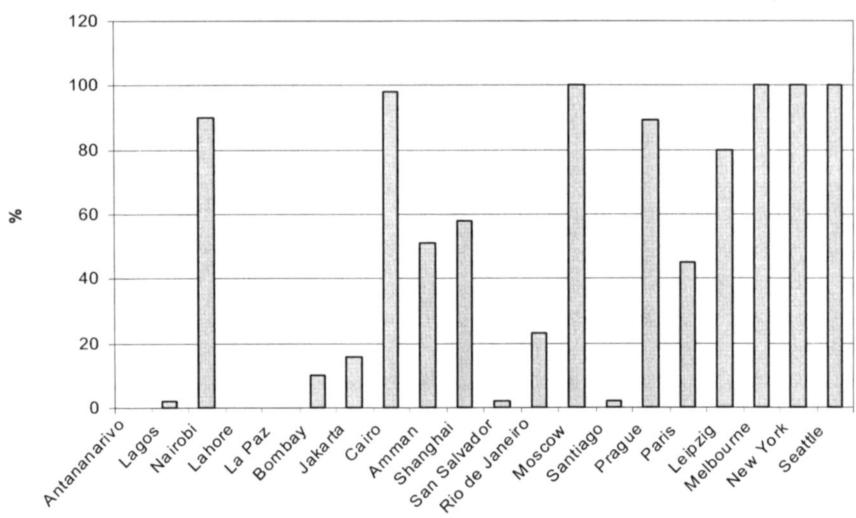

10.5 *Percentage of Wastewater Treated in Selected Cities, 1993.*
Source: Graphed from data courtesy of the World Resources Institute *et al.* (1998) Table 9.3, pp. 278–9.

Phnom Penh and Hanoi have rates of unaccounted for water losses in excess of 60 percent whereas the figure for Singapore is only six percent (Asian Development Bank 1997). These percentages are shown in Figure 10.6. As with inefficiencies in the electricity grids of many developing countries, new devices based on IT could be used to improve the efficiency of water distribution systems. As discussed in more detail in Chapter 5 of this volume, sensors and other monitoring systems can be used to locate important leakages and crack down on illicit connections.

Similarly, monitoring networks could be deployed to learn more about water quality. This is of particular importance in developing countries given the high economic and social costs associated with water-borne diseases. Such health outcomes are caused by pathogens and pollution and relate to lack of access to safe water. In Bangladesh, for example, recent efforts to change the supply of water from contaminated surface waters to cleaner water from aquifers met with the unexpected outcome of arsenic poisoning (World Bank 2003: 173). Such examples suggest the need for greater water testing and water quality monitoring in developing countries to ensure safe water supply.

An area where IT could have an important impact is in encouraging stakeholder participation in the management of water pollution and water quality. In recent years, information databases available to the public through the Internet that respond to mandatory reporting requirements have provided information about pollutant emissions to residents, environmentalists and other interested parties. These systems are known as Pollutant Release and Transfer Registers (PRTRs). The goal is to encourage polluters to monitor their activities and to attempt to reduce pollutant emissions (OECD 1999). Public access to this information also encourages stakeholders to pressure polluters to pollute less.

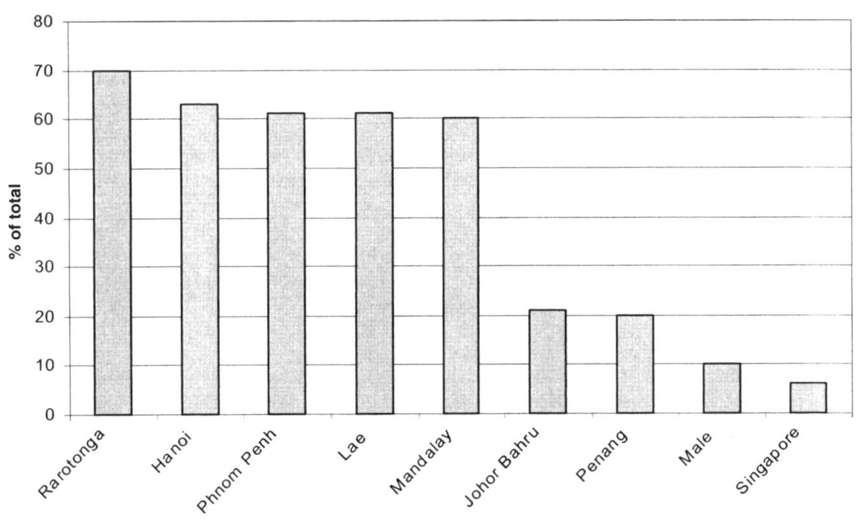

10.6 *Unaccounted for Water Losses in Selected Asian Cities.*
Source: Graphed from data courtesy of the Asian Development Bank (1997).

An example is the Toxics Release Inventory (TRI) in the United States. This is a database maintained by the US Environmental Protection Agency (EPA) that provides information about toxic chemical releases and waste management activities by a number of industry groups and government facilities (USEPA 2003). This electronic database is available on the Internet and allows users to find information about the release of toxic chemicals in their neighborhood by giving them the option of entering their postal zip code and obtaining a report.

Water quality management efforts based on this kind of system are also emerging in developing countries. In Indonesia, the Environmental Impact Management Agency (BAPEDAL) implemented the Program for Pollution Control, Evaluation, and Rating (PROPER) in order to encourage medium- and large-scale polluting firms in river basins to reduce water pollution by way of public disclosure of information and community participation. Under the PROPER program, data about pollutant emissions is collected and verified from participating factories. The data is then analyzed and a color rating is given based on the factory's compliance with water quality regulations. The ratings are then made available to the public. The PROPER ratings include five color categories. *Gold* is given to factories that use clean technology and practice waste minimization and pollution prevention. *Green* indicates good plant maintenance and environmental performance above standards. A *blue* rating means the plant is making efforts to meet minimum standards. *Red* suggests current environmental management efforts do not meet the standards. And a *black* rating means a plant does not implement any pollution control efforts, and is responsible for serious environmental damage. When the program was initiated in 1995 with a group of 187 factories, 65 of them were not in compliance with water quality standards. After 18 months of implementation, PROPER reduced pollution by approximately 40 percent in participating factories (World Bank 2000b). A similar program titled "Industry Ecology Watch" was introduced by the Department of Environment and Natural Resources (DENR) in the Philippines in 1997 (Sandique 1996).

The potential of these kinds of information-based systems to improve environmental performance was recognized by the Organisation for Economic Co-operation and Development (OECD) in 1996 when it held the International Conference on Pollutant Release and Transfer Registers (PRTRs). The OECD also produced a guidance manual for the design and implementation of these systems (OECD 1999). The use of pollutant release and transfer registers with public disclosure facilitated by IT represents an important opportunity for developing countries to improve their information systems about water quality and pollution and to encourage the participation of various stakeholders such as non-government organizations (NGOs), the private sector and community groups in water resources management.

Transportation

The rapidly growing populations of urban areas around the developing world are likely to pose serious challenges to transportation authorities in those countries. Already, cities such as Bangkok, Mexico City, Sao Paulo and many others face important traffic congestion problems. These problems are also present in developed countries. Although the ratio of cars to people is still much higher in developed countries than in developing countries, car ownership is growing more rapidly in the latter group of countries. Figure 10.7 shows the number of cars per 1,000 people in selected cities around the world. Seattle, which is representative of auto intensive cities in developed countries, has more than 600 cars per 1,000 people. The figures for developing country cities such as Bombay, Cairo, Shanghai and Jakarta are well below 100 (World Resources Institute *et al.* 1998: 278–9). In spite of this current pattern, the greatest increase in vehicle ownership in the next few decades is expected to take place in developing countries, especially in East Asia (World Resources Institute *et al.* 1996).

In terms of transportation infrastructure, the data reported shows there can be large variations even among developing countries. In terms of roads paved as a percentage of total roads, countries such as Jordan and Thailand report very high rates whereas countries such as Kenya, Bolivia, Brazil and Chile had rates below 15 percent (World Bank 1999: 264–5). These percentages are shown in Figure 10.8. Such differences are likely to reflect differences in the geography of each country. Even among the developed countries, the United States and Australia have a significantly lower percentage of their roads paved compared to some countries in

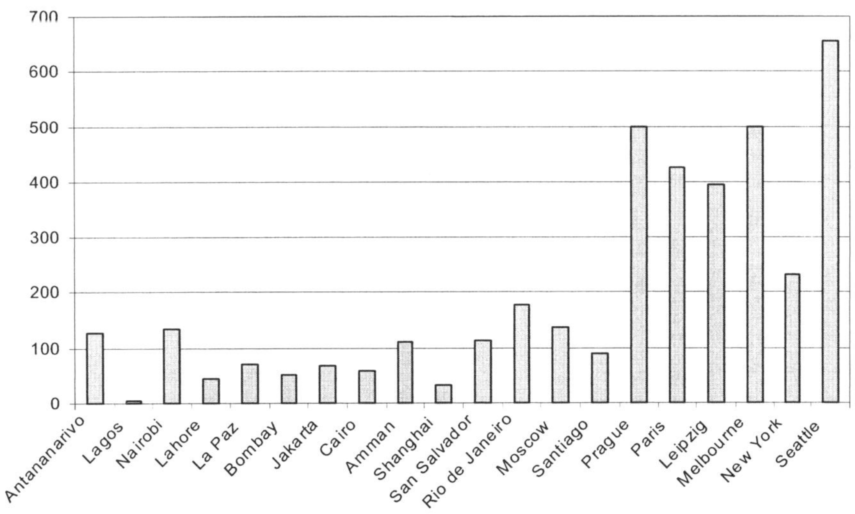

10.7 *Cars per 1,000 Population in Selected Cities Around the World.*
Source: Graphed from data courtesy of the World Resources Institute *et al.* (1998) Table 9.3, pp. 278–9.

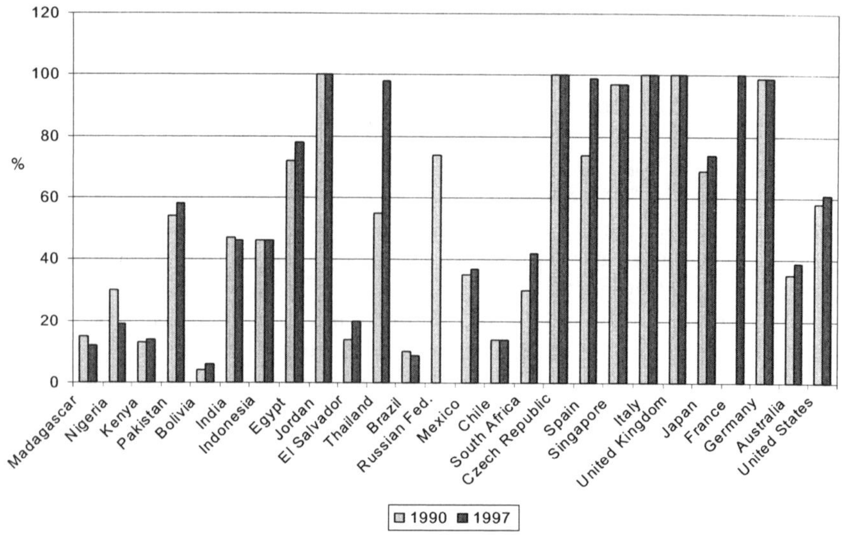

10.8 *Percentage of Paved Roads.*
Source: Graphed from data courtesy of the World Bank (1999) Table 18, pp. 264–5.

Europe. As population, urban areas and car ownership increase in developing countries, their transportation infrastructure is also likely to grow rapidly.

In the area of transportation there are a number of IT systems currently being tested and used in cities in developed countries that could be adapted to conditions in developing countries to alleviate congestion and improve service. Some of the technologies that make these IT-based systems possible are described in Chapter 6 of this volume. These efforts are occurring on an international scale as well. Singapore has been a pioneer in this area. For example, a network of private taxi cabs is equipped with geopositioning systems (GPS) that allow transportation authorities to track the speed of traffic throughout the city. Cameras connected to the front of buses and linked by wireless technology to local bus depots record the license plates of vehicles that use bus-only lanes. Electronic boards controlled wirelessly provide information about delays or accidents. Buried inductive loops collect traffic density data and transmit it to local authorities. Web pages are used to provide information to users about traffic and estimated travel times. And sensors at important intersections can tell how many vehicles are waiting at a light and adjust the time of the lights accordingly (Baum 2001; Baum and Schmidt 2001).

The development of intelligent transportation systems based on IT has also allowed for the application of differential road pricing schemes that change depending on the time of day and aim to reduce congestion. Singapore, Oslo and London have implemented such programs and many other cities around the world are considering similar programs. The rapid development and successful application of intelligent transportation systems in developed countries could find important applications in the urban mega-cities of the developing world. The city of London, for example, began to implement the largest congestion pricing scheme in the world in February 2003, and so far the program has been very successful. Vehicles that enter an area of central London between 7 am and 6:30 pm on weekdays are charged a flat fee electronically using transponders. Six months after the program was introduced, congestion in the charge zone was reduced by 30 percent, the number of vehicles entering the zone decreased by 16 percent and no displaced traffic was observed in other parts of the city. In addition, many people switched from using private cars to public transportation, car pooling and other forms of transportation (City Mayors 2003).

Another area where information technology may find important applications in developing countries is in supporting freight transportation and the movement of goods. Singapore, Taiwan and Hong Kong have adopted integrated transport logistics systems that rely on IT to provide services such as sourcing, inventory management, and automated warehousing and product distribution. In addition, these countries have implemented electronic data interchange systems to facilitate cargo movement through their ports. Systems referred to as value added network services (VANS) link agents participating in the trading of goods such as shippers, freight forwarders, couriers, and cargo distributors through an electronic information system that facilitates making customs declarations and documents, and the exchange of trading documents (Mody and Reinfeld 1997).

Finally, remote sensing represents an important IT application for the purpose of evaluating the role that transportation emissions, as well as emissions from electric power and other stationary sources, play in air quality. Considerable progress has been made in this area in recent years. In addition to its use for monitoring the transport of air pollutants at the global scale, remote sensing is also useful for monitoring sources of emissions at a local level. Many developing countries have vehicle fleets that are relatively old and hence emissions per vehicle tend to be high. As countries begin to implement controls on vehicular emissions in order to improve air quality in their urban environments, they could use on-road, remote sensing technologies to detect vehicles with high emissions of carbon monoxide (CO), nitrogen oxides (NOx) and hydrocarbons and require the owners of those vehicles to ensure adequate maintenance (Science Applications International Corporation 2002: 27).

Energy and Electricity

Differences in energy consumption and electricity use around the world show a similar pattern to that of income per capita. Figure 10.9 shows commercial energy use per capita in kilograms of oil equivalent for a number of countries. At the high end of the scale, the United States had a per capita energy consumption equivalent to 8,051 kilograms of oil equivalent in 1996. In countries such as Kenya, Pakistan, India and Bolivia the equivalent figure was less than 500.

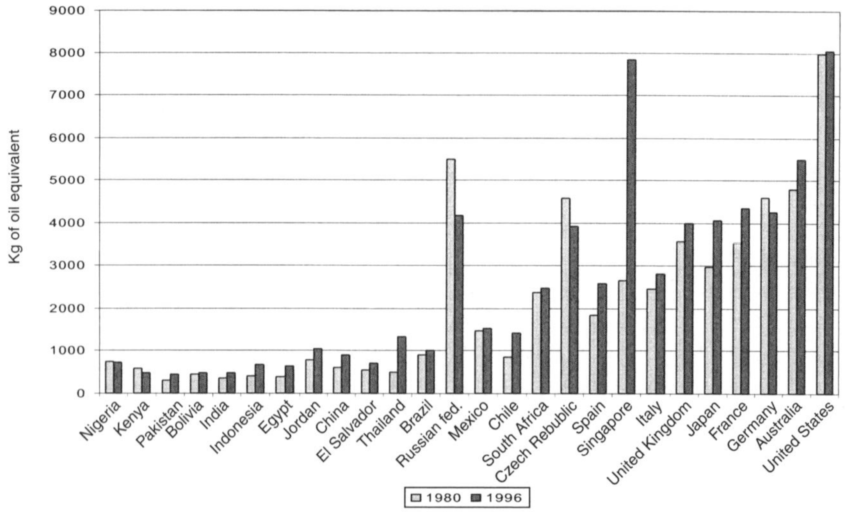

10.9 *Commercial Energy Use Per Capita.*
Source: Graphed from data courtesy of the World Bank (1999) Table 10, pp. 248–9.

Although the figures grew for many developing countries between 1980 and 1996, they were still well below those of the developed countries in the mid-1990s. Given its rapid economic growth, Singapore experienced a remarkable growth in its energy consumption per capita during this period. The opposite is true of the Russian Federation which underwent important changes in its economy and energy sector in the 1990s (World Bank 1999: 248–9).

A similar pattern is observed for electric power consumption. Figure 10.10 shows electric power consumption per capita in kilowatt-hours. Electric power consumption is a measure of electricity production in power plants and combined heat and power plants minus distribution losses and the internal consumption of these facilities (World Bank 1999: 244). Again, the United States has by far the highest electricity consumption per capita of the countries shown. The figure in 1996 was 11,796 kilowatt-hours per person. Countries such as Indonesia, Kenya and Nigeria consumed less than 300 kilowatt-hours per capita in the same year. With the exception of the Russian Federation, the countries shown in Figure 10.10 experienced significant increases in their consumption of electric power per capita during the period 1980–1996 (World Bank 1999: 264–5). As population grows and economies expand, demand for electric power in many developing countries is likely to grow rapidly.

In terms of efficiency in electricity systems around the world, developing countries seem to be at a loss; their grids have much higher transmission and distribution losses as a percentage of total output than those in developed countries. The figures for a sample of countries are shown in Figure 10.11. In Nigeria these losses amounted to over 30 percent of the total in 1996, and Kenya, Pakistan and India had losses greater than 15 percent of total output. At the other end of the

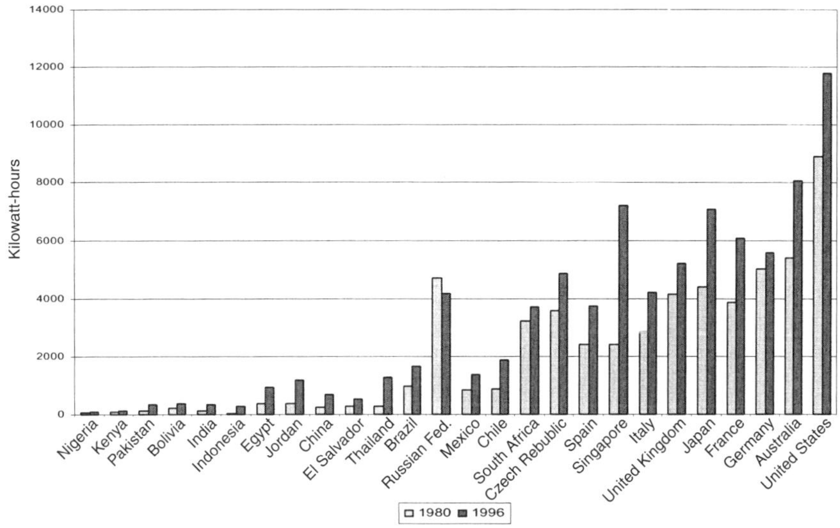

10.10 *Electric Power Consumption Per Capita.*
Source: Graphed from data courtesy of the World Bank (1999) Table 18, pp. 264–5.

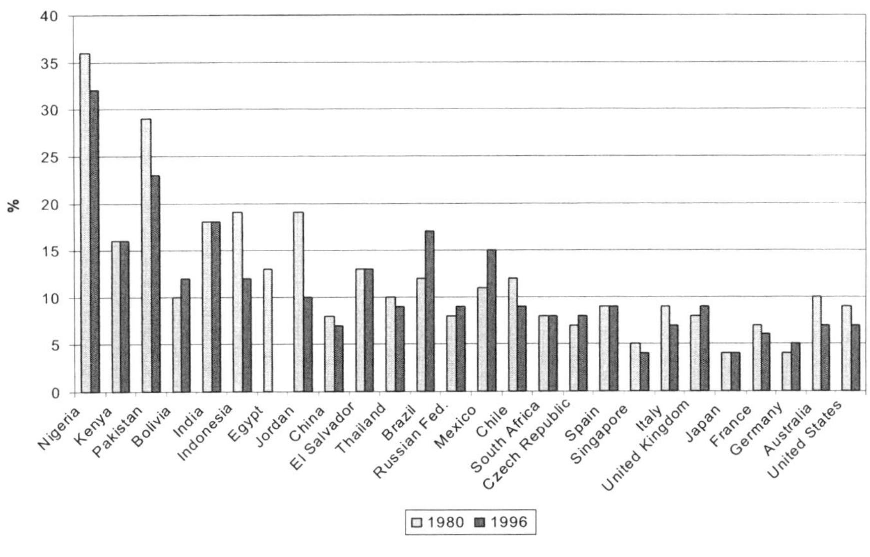

10.11 *Transmission and Distribution Losses in Electricity Grids as Percentage of Output.*
Source: Graphed from data courtesy of the World Bank (1999) Table 18, pp. 264–5.

scale, Japan and Singapore had losses as small as ten percent. Although in most of the countries shown these losses were reduced between 1980 and 1996, in Mexico and Brazil they increased (World Bank 1999: 264–5).

Emerging energy production and distribution methods based on information technologies may be of great benefit to developing countries in the upcoming decades. Smart energy networks of the future will likely consist of a diversified set of energy resources located closer to consumers. New decentralized sources of cleaner power would benefit developing countries that already suffer from inefficient and dirty energy production methods, and may one day replace large-scale energy projects, such as mega hydro-electric dams that have displaced thousands of people and created many environmental problems. These new networks will rely on sensors, solid-state controllers and flexible AC transmission systems throughout the supply chain to create interactive energy markets. These technologies will allow for distributed generation based on technologies such as photovoltaics, wind power, and fuel cells. Intelligent electricity networks will also allow consumers to know pricing conditions in real-time so they can adapt their consumption patterns accordingly (Silberman 2001).

As developing countries attempt to meet the demands of their growing populations and economies they could utilize some of these technologies in order to improve efficiency in their energy systems and expand access to areas where utility companies are currently absent or unable to access. One way to keep abreast of technological developments in this area is through collaborative research information sharing efforts. An example of such an effort is the Energy Technology Data Exchange (ETDE), a program of the International Energy Agency (IEA). Access to ETDE's database is by country membership and the benefits of participation include access to the latest research developments and participation in collaborative projects in energy and economic development projects (ETDE 2003). Participating countries in the ETDE include Brazil, Canada, Denmark, Finland, France, Germany, Italy, the Republic of Korea, Mexico, The Netherlands, Norway, Spain, Sweden, Switzerland, the United Kingdom, and the United States. Countries such as China, Egypt, Hungary, India, and South Africa are potential members (USDOE 1997).

In Mexico, the Institute of Electric Research, founded in 1975 to carry out research in the energy and electricity sector, is already playing an important role in the analysis and dissemination of information about the application of IT to electricity infrastructure. They recognize that in the future environment of energy services and markets that are resulting from deregulation in the sector, integrated systems of real-time information will be key tools to improve efficiency and to reduce the vulnerability of electricity generation, transmission and distribution (Villavicencio 2002). In addition, the Institute has developed the Automated Energy Acquisition System for one of its main clients, the Light and Power of the Center Company. The latter provides electricity to 20 million customers in the central part of Mexico. This system allows for the automated gathering of information about energy throughout the generation, transmission and distribution systems and allows the company to evaluate energy losses and associated costs (Uribe *et al.*

2002). The Light and Power of the Center Company is also automating its substations and integrating information systems that allow it to operate its substations through local or remote supervision and control using Digital Electronic Instruments (DEI), intelligent measuring devices, and wired information systems through LAN and WAN networks (Picasso and Astorga 2003).

Telecommunications and Information Technology

The difference in use of telecommunications and information technology infrastructure between developed and developing countries is perhaps the most dramatic of any of the infrastructure systems considered here. This is not surprising given that this is probably the sector where technology is evolving most rapidly. Figure 10.12 shows differences in telephone lines per 1,000 people in 1997 for a number of countries. The developed countries towards the right side of the graph have over 400 lines per 1,000 persons. The highest number is 644 for the United States. The poorest countries in the sample, like Madagascar, Nigeria, Kenya and India have less than 20 lines per 1,000 persons (World Bank 1999: 266–7).

A similar pattern is observed for the use of mobile telephones. Figure 10.13 shows the variation for the same sample of countries. At the high end of the scale, countries such as Japan, Singapore and Australia had over 250 mobile phones per 1,000 persons in 1997, whereas several developing countries reported having less than 5 per 1,000 persons (World Bank 1999: 266–7). It should be noted that there are important differences among developing countries in mobile phone

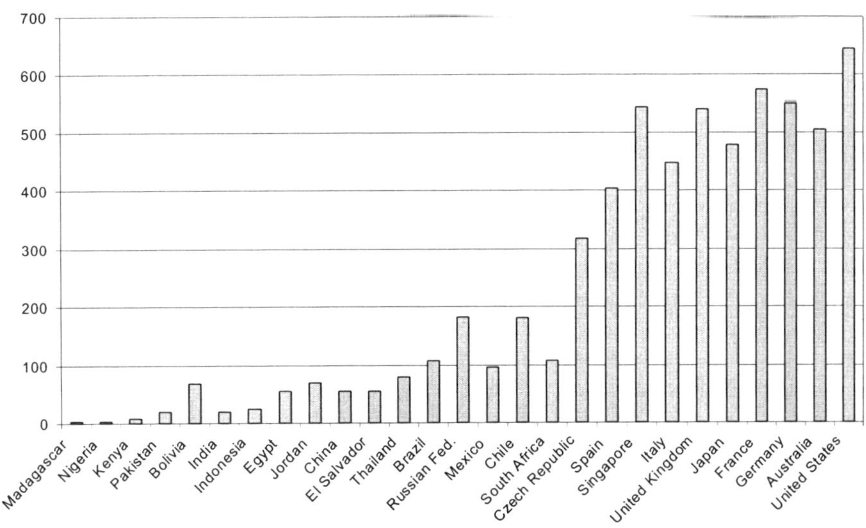

10.12 *Telephone Lines per 1,000 Persons, 1997.*
Source: Graphed from data courtesy of the World Bank (1999) Table 19, pp. 266–7.

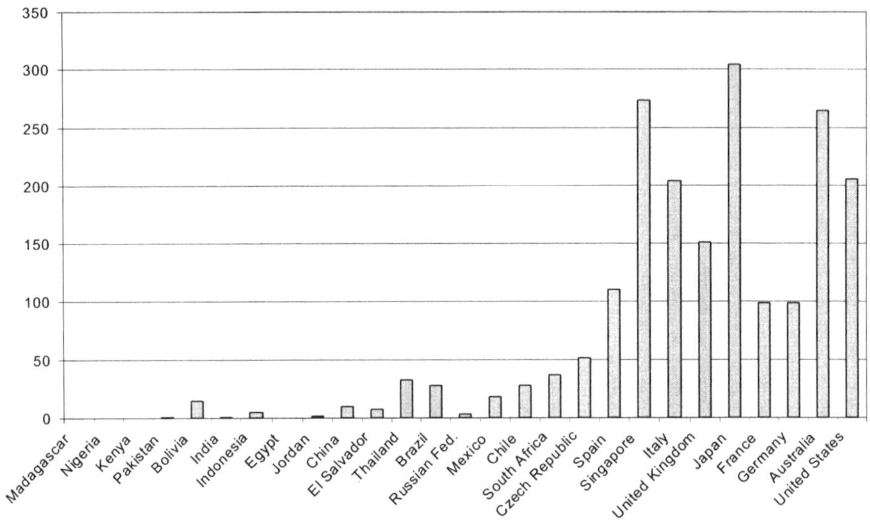

10.13 *Mobile Telephones per 1,000 Persons, 1997.*
Source: Graphed from data courtesy of the World Bank (1999) Table 19, pp. 266–7.

penetration. For example, Thailand has more mobile phones than all of Africa, excluding South Africa (Alaedini and Marcotullio 2002: 93). Despite such differences, mobile telephones have had an important impact in many developing countries because they have given many people access to personal communication in areas where utilities did not extend the traditional telephone services (World Bank 1999: 147). In fact, the ratio of mobile phones to traditional phone lines is higher in some middle-income countries such as The Philippines than it is in developed countries such as the United Kingdom or the United States (World Bank 1999: 57). In Peru there were 1.1 million users of mobile telephones and 1.68 million users of fixed lines in 2000. In that year, about seven percent of households living in extreme poverty had mobile telephones (Fernández-Maldonado 2003).

The rapid adoption of mobile phones in some developing countries shows that a deficit in a traditional infrastructure sector can potentially be leapfrogged by the development of new wireless information technologies. In Haiti, which has a poorly developed telephone infrastructure, wireless technology is used to provide Internet and e-mail access even in rural areas (Peha 1999). The success of these technologies, however, is often closely tied to the regulatory environment in which they are introduced. For example, in La Paz and El Alto, two cities in Bolivia, the licensing of a second provider of mobile telephone services resulted in a more competitive market. Prices decreased and additional services such as prepayment cards were introduced. The result was a ten-fold increase in mobile telephone penetration during the period 1996–1999 and the number of mobile phones surpassed fixed telephone lines by 40 percent (Foster and Irusta 2003).

As with car ownership, access to personal computers in developing coun-

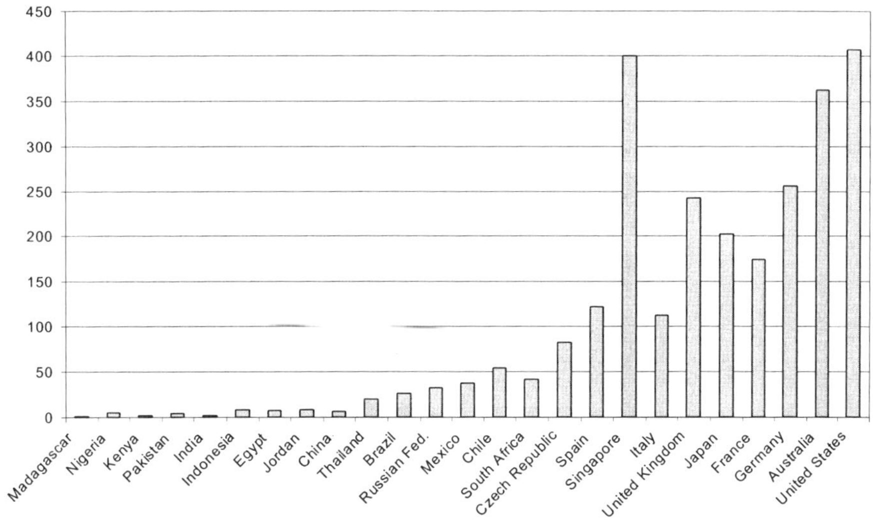

10.14 *Personal Computers per 1,000 Persons, 1997.*
Source: Graphed from data courtesy of the World Bank (1999) Table 19, pp. 266–7.

tries is much more difficult than in developed countries. In 1997, Singapore and the United States had about 400 personal computers per 1,000 persons. In many of the developing countries shown in Figure 10.14, such as Madagascar, Nigeria, Kenya, Pakistan, India, Indonesia, Egypt, Jordan and China, the figure was less than ten (World Bank 1999: 266–7). The disparities in Internet hosts per 10,000 persons in 1997, shown in Figure 10.15, were even more dramatic than the figures for computer ownership. The figures for Madagascar, Nigeria and India were 0.04, 0.03 and 0.13 respectively. The United States had 1,131 Internet hosts per 10,000 persons in 1997 (World Bank 1999: 266–7).

Although these differences in computer ownership and Internet hosts are dramatic, there is evidence that these resources are used more intensively in developing countries. Estimates from 2001 indicate that in Peru there were 35 times more Internet users per Internet host than in the United States. This is attributed to the proliferation of small cyber café style operations called "cabinas públicas de internet." These small entrepreneurial efforts have sprung up all over Lima, the capital city, even in very low-income neighborhoods. They provide services such as access to e-mail and the World Wide Web, voice through the Internet, faxing, scanning, video-conferencing, printing, photocopying, and CD writing. Approximately 40 percent of Internet users in Lima are low-income users. Most of these users find it prohibitively expensive to purchase a computer or to set up a private Internet connection at home, but have access to these services in the cabinas públicas, which price their services at a level that make them affordable to most people (Fernández-Maldonado 2003).

Figures for Internet users per 1,000 persons are shown in Figure 10.16 for

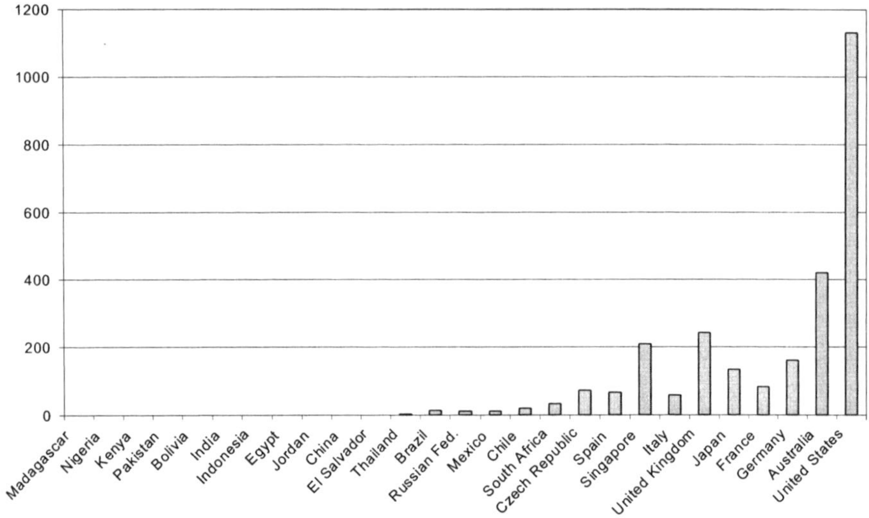

10.15 *Internet Hosts per 10,000 Persons in January, 1999.*
Source: Graphed from data courtesy of the World Bank (1999) Table 19, pp. 266–7.

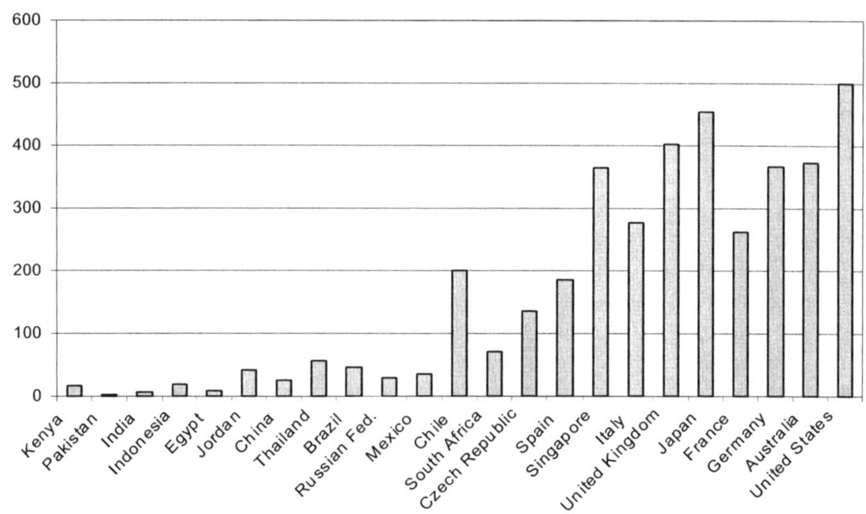

10.16 *Internet Users per 1,000 Persons, 2001.*
Source: Graphed from data courtesy of the World Resources Institute *et al.* (2003) Data Table 1, pp. 234–5.

the year 2001. They show a similar pattern to that of Figure 10.15. In the United Kingdom, Japan and the United States the figures are 403, 455 and 500 respectively. The figures drop to 46, 35 and 30 for Brazil, Mexico and the Russian Federation. At the low end of the scale for this sample of countries, Pakistan, India and Egypt had about 3, 7 and 9 Internet users per 1,000 persons in 2001. The aggregate figure for developing countries in that year was 26 Internet users per 1,000 persons and for the developed countries it was 286 or eleven times greater (World Resources Institute *et al.* 2003: 234–5). Another characteristic of Internet use in developing countries is the disparity in access between urban and rural areas. In countries such as Kenya, Russian Federation and Argentina, about 85 percent, 64 percent and 60 percent of all Internet users were located in the capital city at the end of the 1990s (Petrazzini and Kibati 1999).

Given the importance of information in a global economy, this difference in access to communications and information services presents a very important challenge to developing countries, especially due to the growing importance of IT to infrastructure service provision. The successful adoption of information technology depends not only on income levels but also on access to other human capital skills such as education and technical training. Some authors have noted that lack of access to IT may widen the income gap between developed and developing countries, and may exacerbate income differences between urban and rural areas in many developing countries (Alaedini and Marcotullio 2002: 89). Realizing the importance of this sector to their future economic growth and social well-being, many developing countries are adopting specific policies to address the issue of improving access to these services. Malaysia, for example, has included IT as a central tenet of its national development strategy. Among the stated goals of this strategy are improving awareness of new technologies among its population, facilitating the diffusion of these technologies, and expanding education and training in these areas (World Bank 1999: 61). In addition, several countries in Africa are developing national plans and strategies to develop their IT and communications infrastructures. The National Information and Communication Infrastructure (NICI) initiative is a process that brings together stakeholders in countries across Africa to promote IT. In several countries, including Egypt, Mauritania, Morocco, Nigeria, Rwanda, Senegal and Tunisia, the government has established a ministry to address the development of IT (GKP 2002).

The global divide in access to information technology has sparked efforts to better understand and bridge this divide, and to use IT to promote sustainable development. One example is the Digital Opportunity Initiative, a public-private partnership formed by Accenture, the Markle Foundation and the United Nations Development Programme (UNDP). In their report titled *Creating a Development Dynamic*, they provide numerous examples of how IT is being adopted and how it has improved lives in different developing country settings. In the area of health, a service based on e-mail and Internet called HealthNet allows health care users from developing countries to access medical information and databases, to participate in physician collaborations, and to share information about research efforts and areas of common interest. In the area of education, IT has found important applications in the area of distance learning in higher education. Countries such as Turkey, Indonesia, China, India, Thailand and Korea have the largest distance-learning universities in the world. The Internet is also providing many opportunities to share information about research. The African Virtual University encourages shared research by providing chat rooms, e-mail, bulletin boards and home pages. In the area of communications, the Grameen Bank facilitates small loans to women in Bangladesh to purchase cellular phones. Owners of these phones then rent them out to other users. Access to communications technology facilitates access to market information about agricultural goods and inputs, weather reports, and disaster alerts. Similarly, access to the Internet allowed vegetable producers in Chincheros, a small village in Peru, to form a partnership with an export company and sell their produce in New York (Accenture, Markle Foundation and UNDP 2001).

Another global effort to promote widespread access to IT around the world as a tool to foster sustainable development and improve people's lives is the Global Knowledge Partnership (GKP). This partnership is made up of private, public and non-profit organizations from countries around the world. It was created after the 1997 Global Knowledge Conference, which was sponsored by the World Bank and the Government of Canada. A follow-up conference was held in Malaysia in 2000 and representatives from about 75 percent of the developing world attended. The Secretariat of GKP is currently in Malaysia (GKP 2002). About 80 organizations from 38 countries are currently members of this partnership (GKP 2003).

GKP has also documented a large number of success stories of the use of IT to improve people's lives. One example is the Digital Divide Data (DDD) project in Cambodia. It integrates education, vocational training and work and is aimed at disadvantaged youth. Employees at DDD receive images from clients around the world and convert them to ASCII files, create databases, and/or add HTML and then e-mail the output. Another example is the Engineers Without Borders Scala Project in the Philippines. This project was started by youth in Canada as an effort to help Filipino youth gain greater access to IT and aims to train 3,000 young people every year in IT skills through programs delivered through a network of 19 information and communications technology centers (GKP 2003). An additional example is the Acacia initiative in Uganda, which provides access to Internet, telephone, fax and photocopying services to people in rural areas through "telecenters" (GKP 2002).

Additional partnerships to better understand the global impact of IT are also being formed as a result of the World Summit on the Information Society (WSIS), organized by the International Telecommunications Union (ITU) and the United Nations. The goal of the summit is to gather key stakeholders from United Nations agencies, member state governments, the private sector and civil society to develop a declaration of principles and a plan of action to facilitate the creation of a global Information Society. The first part of this two-part summit took place in Geneva in December, 2003 (WSIS 2003a). The plan of action agreed to includes a number of targets for the year 2015 that include using information and communications technology to connect: villages; universities, colleges, secondary schools and primary schools; scientific and research centers; public libraries, cultural centers, museums, post offices and archives; health centers and hospitals; and local and central government departments. Additional goals include: adapting all primary and secondary school curricula to meet the challenges of the Information Society; ensuring that all of the world's population have access to television and radio services; facilitating the presence and use of all world languages on the Internet; and ensuring that more than half the world's inhabitants have access to IT and communications technology within their reach (WSIS 2003b).

Conclusions

Current population projections indicate that by the year 2015 most of the large urban agglomerations of the world will be in developing countries. In addition, while many developed countries are beginning to experience zero or negative population growth rates, many developing countries will continue to observe significant growth rates through the middle of this century. Many developing countries are already coping with serious infrastructure deficits and their rapidly growing urban areas will no doubt present them with additional challenges. Although applications of IT to infrastructure systems are currently being developed and deployed more commonly in developed countries, in the future developing countries are likely to benefit from them.

In many cases, emerging technologies based on IT may allow developing countries to leapfrog traditional barriers to accessing and improving infrastructure services. For example, the deficit in the provision of telephone main lines has been reduced in many developing countries due to the availability of mobile phones. In Haiti, a country with only 0.9 telephone lines per 100 people, three out of four Internet service providers were able to use wireless technologies to offer Internet access even in rural areas (Peha 1999). There are examples from many countries where access to mobile telephone services and the Internet in poor and remote rural areas has improved lives and expanded economic opportunities. The Gyandoot project in Madhya Pradesh, a poor area of India, has successfully established computer kiosks that provide information about popular crop prices so that farmers know when to sell. The kiosks use content developed specifically to service the economic and social needs of the people they serve (Warschauer 2003). Because of the potential of IT to improve people's lives, many countries have designed specific policies and strategies to expand access to communications and IT services. In addition, several partnerships and international collaborations, such as the Global Knowledge Partnership and the Digital Opportunities Initiative, aim to bridge the global digital divide.

In traditional infrastructure areas such as transportation, energy, and water and wastewater treatment, IT offers developing countries the opportunity to improve efficiency and management. In the area of transportation IT has successfully contributed to mitigating traffic congestion and reducing emissions in several cities in developed countries. The lessons from these experiences could be adapted to heavily congested and polluted cities in developing countries. In the area of energy and electricity IT is already promoting research collaboration and technology diffusion through efforts such as the Energy Technology Data Exchange (ETDE), a program of the International Energy Agency (IEA). Research institutions in developing countries, such as the Institute of Electric Research in Mexico, are already implementing IT to improve management and reduce energy losses in their electric systems. Similarly, in the area of water, IT can be used to address the issue of water losses that tend to be high in many cities in developing countries. IT is already being used in information-based systems to improve water quality and reduce pollution. Pollutant Release and Transfer Registers (PRTRs) programs have been successfully implemented in Indonesia and the Philippines and are being developed in other countries. These programs can reduce important information gaps about water pollution and encourage the participation of various stakeholders in the management of water resources.

Despite the promise of IT to solve many of the infrastructure problems in developing countries, it is important to note that the successful application of such technologies will depend on sustained local financing as well as local regulatory support, managerial commitment and cultural acceptance. Such factors may be much different from those found in cities across developed countries. A regulatory environment that fosters competition may be crucial to facilitate the diffusion of IT in developing countries. The expansion of mobile phone penetration in two Bolivian cities following the introduction of competition described earlier illustrates this point (Foster and Irusta 2003). The introduction of IT is most successful when it is done by addressing the complex economic and social issues of the environment where it is introduced (Warschauer 2003). An important research agenda in developing countries is how to adopt and use IT-based infrastructure technologies as countries deregulate infrastructure services, encourage private sector participation and decentralize service provisions, as these are some of the policy trends that are currently being adopted in many developing countries (World Bank 2003).

References

Accenture, Markle Foundation and United Nations Development Programme (UNDP) (2001) *Creating a Development Dynamic: Final Report of the Digital Opportunities Initiative*, Online. Available at: http://www.opt-init.org/framework.html (accessed December 30, 2003).

Alaedini, P. and Marcotullio, P. (2002) "Urban Implications of Information Technology/New Electronics for Developing Countries," *Journal of Urban Technology*, Vol. 9, No. 3, pp. 89–108.

Asian Development Bank (ADB) (1997) "Asia Has More Safe Water – But Still Wastes 35 Percent, ADB Data Book Shows," Online. Available at: http://www.adb.org/Documents/News/1997/nr1997111.asp (accessed January 8, 2004).

Baum, D. (2001) "The Perfect Commute," *Wired*, Vol. 9. No. 11, pp. 170–81.

Baum, D. and Schmidt, S. (2001) "Singapore: The State As Traffic Cop," *Wired*, Vol. 9, No. 11, pp. 175–6.

City Mayors (2003) "London's congestion charge cuts traffic jams by 30 per cent," Online. Available at: http://www.citymayors.com/report/congestion_charge.html (accessed December 29, 2003).

Energy Technology Data Exchange (ETDE) (2003) "About ETDE," Online. Available at: http://www.etde.org/ (accessed December 30, 2003).

Fernández-Maldonado, A.M. (2003) "The Diffusion and Use of Information and Communications Technologies in Lima, Peru," *Journal of Urban Technology*, Vol. 8, No. 3, pp. 21–43.

Foster, V. and Irusta, O. (2003) "Does Infrastructure Reform Work for the Poor? A Case Study of the Cities of La Paz and El Alto in Bolivia," *World Bank Policy Research Working Paper 3177*, Online. Available at: http://econ.worldbank.org/files/31791_wps3177.pdf (accessed January 6, 2004).

Global Knowledge Partnership (GKP) (2002) *GKP Recommendations on Issues of Bridging the Digital Divide*, Kuala Lumpur: Global Knowledge Partnership Secretariat, Online. Available at: http://www.globalknowledge.org/ (accessed December 30, 2003).

Global Knowledge Partnership (GKP) (2003) *Youth. Poverty. Gender. ICT for Development Success Stories*, Kuala Lumpur: Global Knowledge Partnership Secretariat, Online. Available at: http://www.globalknowledge.org/ (accessed December 30, 2003).

Mody, A. and Reinfeld, W. (1997) *Infrastructure Strategies in East Asia: The Untold Story*, in Mody, A. (ed.), Washington, D.C.: The World Bank, pp. 131–46.

Organisation for Economic Co-operation and Development (OECD) (1999) *Proceedings of the OECD International Conference on Pollutant Release and Transfer Registers (PRTRs)*, Part 1, Paris: OECD, Online. Available at: http://www.olis.oecd.org/olis/1999doc.nsf/LinkTo/env-jm-mono(99)16-part1 (accessed December 30, 2003).

Peha, J.M. (1999) "Lessons from Haiti's Internet Development," *Communications of the ACM*, Vol. 42, No. 6, pp. 67–72.

Petrazzini, B. and Kibati, M. (1999) "The Internet in Developing Countries," *Communications Of The ACM*, June, Vol. 42, No. 6, pp. 31–6.

Picasso, C. and Astorga, C. (2003) "Sistema de control supervisorio basado en PC aplicado en automatización de subestaciones y centros de operación regional de redes de distribución," *Boletín IEE*, April–June issue, pp. 68–72.

Sandique, R. (1996) "Rating System for 2,000 Industries in Manila Set," Manila Standard, December 9, Online. Available at: http://www.worldbank.org/nipr/ecowatch/ecowatch3.htm (accessed December 30, 2003).

Science Applications International Corporation (2002) *An Examination of the Application of Remote Sensing Data in Detecting and Characterizing Air Pollution Transport and Emissions*, Research Triangle Park, NC: US EPA, April, Online. Available at: http://www.epa.gov/ttn/chief/eiip/pm25inventory/remsens.pdf.

Silberman, S. (2001) "The Energy Web," *Wired*, Vol. 9, No. 7, pp. 114–27.

United Nations (2003) *World Population Prospects, The 2002 Revision, Highlights*, New York: United Nations Population Division, Online. Available at: http://www.un.org/esa/population/publications/wpp2002/WPP2002-HIGHLIGHTSrev1.PDF (accessed December 29, 2003).

United States Department of Energy (USDOE) (1997) "Mexico Joins Oak Ridge International Information Program," Online. Available at: http://www.oakridge.doe.gov/media_releases/1997/r-97-035.htm (accessed December 30, 2003).

United States Environmental Protection Agency (USEPA) (2003) "Toxics Release Inventory (TRI) Program," Online. Available at: http://www.epa.gov/tri/ (accessed December 30, 2003).

Uribe, C., González, E., Pérez, R, Álvarez, V., Domínguez, N., Carrizosa, V. and Alarcón, E.

(2002) "Sistema de Ayuda para la Implantación del Mercado de Energía en Luz y Fuerza del Centro," *Boletín IEE*, September–October issue, pp. 221–30.
Villavicencio, A. (2002) "Tecnologías en evolución de los Sistemas Integrales de Información en Tiempo Real," *Boletín IEE*, September-October issue, pp. 191–6.
Warschauer, M. (2003) "Demystifying the Digital Divide," *Scientific American*, Vol. 289, No. 2, pp. 42–7.
World Bank (1999) *World Development Report 1998/1999Knowledge for Development*, New York: Oxford University Press.
World Bank (2000a) *World Development Report 1999/2000 Entering the 21st Century*, New York: Oxford University Press.
World Bank (2000b) *Greening Industry New Roles for Communities, Markets, and Governments*, New York: Oxford University Press.
World Bank (2001) *World Development Report 2000/2001 Attacking Poverty*, New York: Oxford University Press.
World Bank (2003) *World Development Report 2004 Making Services Work for Poor People*, Washington, D.C.: The World Bank and Oxford University Press.
World Resources Institute, United Nations Environment Programme, United Nations Development Programme, and The World Bank (1996) *World Resources 1996–97: A Guide to the Global Environment*, New York: Oxford University Press.
World Resources Institute, United Nations Environment Programme, United Nations Development Programme, and World Bank (1998) *World Resources 1998–99: A Guide to the Global Environment*, New York: Oxford University Press.
World Resources Institute, United Nations Environment Programme, United Nations Development Programme, and World Bank (2003) *World Resources 2002–04: Decisions for the Earth: Balance, Voice, and Power*, New York: Oxford University Press.
World Summit on the Information Society (WSIS) (2003a) "Basic Information: About WSIS," Online. Available at: http://www.itu.int/wsis/basic/about.html (accessed January 6, 2004).
World Summit on the Information Society (WSIS) (2003b) Plan of Action. December 12, 2003, Online. Available at: http://www.itu.int/dms_pub/itu-s/md/03/wsis/doc/S03-WSIS-DOC-0005!!PDF-E.pdf (accessed January 6, 2004).

Chapter Eleven

Green Design and Intelligent Buildings

Timothy Beatley

Introduction

Buildings and the built environment in the US – homes, businesses, libraries and schools – consume significant amounts of energy, water and other materials, and thus have major implications for resource management and environmental conservation. Buildings and the built environment account for some 40 percent of our energy needs, produce about one-third of the world's carbon emissions, and take about 40 percent of the raw materials extracted from the earth (City of New York 1999). A more careful design of the built environment that seeks to use energy and water sparingly, recycles materials, minimizes wastes, and enables more sustainable lifestyles represents an important form of urban infrastructure. Increasingly green building and intelligent design take advantage of information technologies (IT), making it easier to reach these environmental goals. This chapter explores the potential of green design and the role of IT at two primary spatial levels – the individual structure or building and the urban district or neighborhood.

Growing Interest in Green Design

Interest in green design and intelligent buildings is certainly not new. Arguably, climate sensitive, sustainable building and settlement designs are as old as civilization itself. An emerging environmental movement and energy shortages in the 1970s generated considerable interest and experimentation, but not until recently, especially in the past five to ten years, has interest in green design and intelligent buildings reached the level of mainstream society.

Several measures or indicators of the growing interest in green design can be cited. For instance, the US Green Building Council's green building certification system, known as Leadership in Energy and Environmental Design (LEED) has experienced increasing popularity. Buildings comprising over 100 million square feet in space had applied for LEED certification by 2003, with the number of buildings growing significantly over time (US Green Building Council undated). New building clients increasingly appear to value the prestige and marketability of this certification, and a growing number of municipalities now either mandate or strongly encourage LEED-certified construction. Cities like Seattle and Portland have adopted municipal policies supporting green design, and other cities such as Austin and Denver have developed local green builder programs to advance and encourage green building among private builders and prospective housing consumers. Efforts to promote green building at the state level can also be seen. The state of California has established a sustainable building task force, and issued an executive order establishing green buildings and design as an important state goal. States such as Massachusetts have established grant programs to encourage the application of green ideas and technologies in the design of schools, as a further example.

Moreover, private companies are increasingly recognizing that green buildings and intelligent design are "good for the bottom line" (Heerwagen 2002; Romm and Browning 1998). A number of recent green corporate office buildings (e.g. GAP headquarters, Herman Miller, Commerzbank, Swiss Reinsurance, among others), show the attraction of conveying commitment to green and sustainable principles.

Benefits of Green Design

The benefits of green buildings and intelligent design are extensive indeed. There is a considerable and growing body of evidence, for instance, that such buildings can substantially improve worker satisfaction and productivity in office buildings. Increases in job satisfaction and worker productivity have been found in green buildings where occupants have direct control over ventilation and temperature (for a review, see Heerwagen 2002).

Incorporation of plants and nature, and the design of buildings to maximize daylight have also been shown to significantly improve occupant and worker satisfaction, as well as yield economic savings (e.g. see Rocky Mountain Institute (RMI) 1998; Romm and Browning 1998). Especially impressive are studies that show the developmental, psychological, and physiological benefits of daylighting especially in schools (e.g. see Beatley (forthcoming)). Day-light schools – schools designed to provide full spectrum daylight in every room – have been shown to result in improved test scores, students that grow faster, and even lowered rates of tooth decay (Heschong Mahone Group 1999; Hathway *et al.* 1992)! In addition, significant reductions in the need for artificial lighting and reduced heating and cooling needs can result. Thus, the economic, health, and environmental benefits of green design and intelligent buildings are considerable indeed.

While green and intelligent features of buildings are often perceived to be too costly, there is considerable evidence that such elements are actually quite cost-effective over the life of the building. A recent review of 33 green buildings in California, for instance, found that while green features added around two percent to the initial cost of these buildings they yielded long-term savings (e.g. in the form of reduced energy consumption) of around 20 percent (Kats 2003). This study provides a strong economic rationale for green design concluding that the "total financial benefits of green buildings are over ten times the average initial investment required to design and construct a green building:"

> Additionally, the relatively large impact of productivity and health gains reflects the fact that the direct and indirect cost of employees is far larger than the cost of construction or energy. Consequently, even small changes in productivity and health translate into large financial benefits. (Kats 2003: ix

Definitions and Concepts

Before progressing further, several preliminary issues need clarifying. In discussing current practice in this area, different terms or versions of terms are employed, and some distinctions between them are useful to keep in mind. *Green* design or *green* buildings incorporate features that in one way or another seek to reduce environmental impacts and effects – whether in the form of reduced energy and water consumption, or promoting the use of environmentally-friendly building materials that have a small or smaller effect on our ecosystems and natural environment, such as recycled carpets and sustainably harvested wood. Green buildings are also typically concerned with creating healthier living and work environments through improved ventilation and use of non-toxic paints and materials that result in better indoor air quality. Of course, green design and building choices often accomplish multiple objectives.

A green building need not necessarily be a very complex structure, and may not necessarily incorporate intelligent design elements (including IT). Designing homes and buildings to capture natural breezes or to incorporate a green rooftop, for instance, may involve few if any truly intelligent features. An *intelligent* build ing tends to connote a more active design configuration, in which elements of the building help it to respond to changing climatic, environmental, user, and occupant circumstances.

Wigginton and Harris (2002: 23) define an intelligent building as "a building with the ability to know its configuration, anticipate the optimum dynamic response to prevailing environmental stimuli, and activate the appropriate physical reaction in a predictable manner." They continue:

> It is expected that the system will strive to exploit the use of natural forces and minimize the need to impact energy from non-renewable sources. The truly "intelligent building" should therefore be endowed with some of the human characteristics that give it the ability to learn, adjust and respond instinctively to its immediate environment in order to provide comfortable internal conditions and use energy more efficiently.

Green and Smart

Green buildings and green design are also often discussed in terms of "high performance" buildings. In 1999, the City of New York adopted a set of "High Performance Building Guidelines," for example, that place considerable emphasis on IT as a means to this high performance. Through its Office of Sustainable Design, a variety of green and intelligent features have been incorporated into new and renovated municipal buildings in that city (City of New York 1999; City of New York 2002).

Intelligent or smart building design, moreover, may address a variety of health, safety and other goals that extend beyond our traditional notions of green architecture. Advanced sensor technology is increasingly viewed as an opportunity for early detection of fire and impending structural failure in the case of, say, a terrorist attack or an earthquake (e.g. Building Design 2003; Flame Retardancy News 2002). Many of the IT applications described here, as a further example, help to facilitate independent living by older members of society. Green design and intelligent design seek to minimize the extent of the ecological footprint, but they embrace and advance many other social goals as well. Such concerns are important, to be sure, but not usually subsumed within green design. Most of what is addressed below falls squarely in the areas of green design, though it should be noted, that IT components of buildings and neighborhoods may have achieved other, non-environmental benefits as their primary goal.

Intelligent buildings certainly can help in significant ways to make structures more green. Similarly, the trend towards green design and building will help push the trend towards intelligent structures, and increasingly creative use of IT. So, intelligent, green and IT go hand-in-hand.

Green Design as Public Infrastructure

These new IT contributions to green design have implications for public infrastructure in several different important ways.

One way involves a reconceptualizing of the very concept of infrastructure itself. As green design and IT elements in individual homes and buildings substantially reduce energy and resource demands, they in turn replace or minimize the need for investments in more conventional forms of infrastructure. If, as in projects like the Greenwich Millennium Village, a new ecological district in London, residents require half or less of the energy typically needed in a conventional building, this translates into the need for a smaller, stronger and more resilient public power grid and less of a need to rely on conventional extensions and expansions of this infrastructure. Buildings like the Herman Miller factory in Michigan, which uses 65 percent less water than a typical building of its size, represent a significant reduction in need for new or expanded water systems and conventional water treatment and distribution of delivery infrastructure. In the same way that power companies and utilities see advantages in energy-efficient and demand management efforts (i.e. fewer power plants that need building or expanding), green and intelligent buildings offer similar benefits of reduced or avoided forms of conventional infrastructure. Onsite stormwater collection and treatment systems, increasingly common in the US and Europe, and the growing interest in green rooftops, have tremendous implication for reduced need for conventional (and expensive) urban stormwater collection systems. The need for heavily engineered (and costly) urban infrastructure like this can be avoided through such green designs. Incorporation of "living machine" technologies – on site wastewater treatment utilizing plants and aquatic organisms – represents both a less environmentally damaging approach to sewage treatment, and again a significant reduction in the need for traditional collection and treatment systems. The Oberlin College Environmental Studies Building, a William McDonough and Partners design, and the brainchild of Program Director David Orr, is an example of a structure that incorporates such technology, treating (naturally) all the wastewater generated by users of the building (see Beatley 2000).

Especially in the area of energy production green design represents a new way of conceiving of the traditional power grid and power production and distribution infrastructure. No longer does energy (or at least not all of it) have to be produced through large, fossil-fuel dependent power generating plants, and transmitted long distances. Every building rooftop, every housing complex, every school, every library, represents the potential for decentralized renewable power production. The Oberlin Environmental Studies building is designed to be a net energy-exporter, for instance, with considerable on-site power production from rooftop solar panels. As a further example, an impressive green structure, the Mont-Cenis Academy, in Herne, Germany, contains 10,000 square meters of photovoltaics panels on its rooftop, producing 1MW of power – much more electricity than is actually needed by the building. An adjacent "energy park" harvests methane gas from abandoned underground mines (the Academy sits on a brownfield site), in turn using it as a fuel for a co-generation plant that produces heat and electricity (Entwicklungsgesellschaft 1998). There are, increasingly, many examples of creatively designing-in energy production, whether in the form of rooftop or façade-mounted photovoltaics, or photovoltaics incorporated directly into roofing tiles, or micro-wind turbines, or renewable-energy biomass-fueled boilers and combined heat and power plants.

IT and Green Buildings: Some Cases

IT is an increasingly common and essential element in the design of new green and intelligent structures. New sensor and monitoring technologies offer substantial opportunities to smarten up buildings even further. Chemical gas sensors and old factory sensors offer new potential for early warning of smoke and fire in buildings (Watson 2002; Flame Retardancy News 2002). More commonly, rooms equipped with light sensors allow automatic adjustment of artificial lighting depending on the amount of natural light entering the structure.

Most large structures now commonly include an integrated building management system (or building automation system) that controls many building functions, including lighting, ventilation, and heating and cooling. These systems allow automatic powering-down of elevators and other systems when not needed, and reduce lighting and heating/cooling when there are no or few occupants in rooms. Such an automated, integrated "nerve system" ties together and controls many things. From adjusting window shades to the opening and closing of windows, IT is an increasingly essential element in creating green, sustainable buildings.

There has been considerable innovation in recent years in designing skyscrapers that reflect new green sensibilities. Many advocates of sustainability now recognize that skyscrapers represent an important part of the solution to global environmental problems, and especially have the potential to achieve much higher levels of energy efficiency, often with significant IT elements. The tallest building in Europe (until recently), the Commerzbank headquarters building in Frankfurt, Germany, is a case in point. Completed in 1997, it represents the best of a new generation of green skyscrapers. This 48-story structure was designed by Norman Foster and Partners, a firm that has been pushing the green design envelope in some impressive ways. The building takes a triangular form, with a long center atrium and a series of nine sky gardens that wind their way up and around the structure. These enclosed gardens are important in providing natural ventilation, in moderating temperatures and in generally enhancing the working conditions in the building. They have become important places for eating, meeting and relaxing.

A number of new energy features are included in this design. A double-skin façade provides an important thermal buffer. Built-in sunshading screens, operable windows in offices, a tremendous amount of daylight (all offices have large windows, many with spectacular views of the River Main and the city skyline), and a ceiling cooling system utilizing chilled water, all reduce the energy demands of this skyscraper. Clearly one of the most noticeable things about the building is the extensive amount of natural daylight in the building. This results from the large interior atrium, the gardens, and the use of glass partition walls throughout. Because the sky gardens snake around the building's façade, natural light finds its way into the interior throughout the course of the day.

Extensive IT features can be found in the building. Motion sensors and light sensors are a key element and serve to significantly moderate and adjust the artificial lighting that is needed. The structure's building management system activates motorized windows and Venetian blinds when necessary to keep the building's temperature within pre-set winter and summer limits. Together these design features make for a very energy-efficient building compared with conventional skyscrapers. The building uses almost 40 percent less energy than a conventional structure, as a result of these many energy features. As well, employees are reported to be very happy with the working conditions of the building.

The new Swiss Re-insurance building, under construction in London, further demonstrates the ecological and energy conserving potential of new large office designs. This 41-story structure, affectionately labeled the "erotic gherkin," takes on the shape of a cigar, tapering at the top and bottom. The winding and tapering shape also maximizes the amount of daylight brought into the building, and a double-layer building skin acts as a thermal buffer, further reducing its energy consumption. The tapered bottom allows more room for public space.

The size and density, along with proximity of structures like the Commerzbank and the Swiss Re-insurance building to existing walkable urban centers, suggest even greater energy savings. While the Commerzbank does include parking, most travel to and from the building happens through public transit. The Kaiserplatz metro station is nearby and the Frankfurt Main Train Station is just a few minutes' walk away.

Extensive digital technologies are also typically utilized in the design process. Parametric modeling and wind tunnel testing have been used to ensure desirable wind flow around the Re-Insurance building, for instance, and to minimize downdrafts experienced at street level. Because the wind moving around the structure will "caress rather than buffet" the structure, the building can actually be lighter (Glancey 2003). The design directs air flow into and through the building through a series of window vents. Six-story high sky gardens spiral around the building, connecting workers, treating the air as it moves through the building and providing important trees and greenspaces for employees.

IT can also be an important active element in maximizing the solar energy potential of green buildings. Buildings like the Heliotrop, in Freiburg, Germany, actually rotate to follow the sun (named after the heliotropic movement in plants). As a further example, the Norman Foster green redesign of the German Reichstag includes as a central feature a rotating sunshade, which tracks the daily movement of the sun, bringing into the building extensive daylight while minimizing glare.

The structure of a greenhouse, in combination with advanced IT, offers a great opportunity to reduce energy consumption, infuse green features and create high quality working and living conditions in modern buildings. The Alterra building, in Wageningen, Netherlands, is one of the most impressive examples of an office structure organized around a greenhouse concept. The building is essentially comprised of two connected greenhouses, with offices, seminar rooms, and a library opening on to these delightfully lush greenhouse interiors. The building uses a standard commercial greenhouse roof, which can be opened and closed depending on the season and time of day. Within these greenhouses are three levels of offices, accessible both from interior hallways and through a series of metal stairs and walkover structures. Uniquely, every office has both a hallway door and a door directly opening onto the gardens. The doors and windows of offices are opened to the greenhouse interior spaces even during winter.

There are many energy conservation advantages to the greenhouse design, and the Alterra Building demonstrates them well. Extensive interior daylight is the first advantage, creating a bright open feeling to the work environment and little need for much artificial lighting. Air circulates and passes through the structure, moving from a cool underground chamber laterally through the building and side glass panels that open automatically. The interior plants and vegetation, and the flowing water moving through the interior ponds and water features, provide evapo-transpiration and natural cooling for the structure. The amount of evapo-transpiration is considerable – an estimated 6,000 liters per day. The building employs a clever water cycle, where stormwater runoff from the roof is sent to a pond system behind the building (really more like a lake) and returned to the interior, filtered, then used for toilet flushing.

It is fascinating to watch how the building automatically adjusts and moderates its mechanical elements to fit changing climate and weather conditions. As the building heats up in the morning, the glass roof and side glass panels gradually open. Watching this occur on a July morning, a slow buzzing of motors can be heard as the roof's shading system unfolds. In a few minutes the interior gardens take on an entirely different look, more like a hidden palm oasis with tree shading and an immediate feeling of coolness. The building's roof can also be closed relatively quickly when rain threatens. The building's automated control system is connected to a nearby weather station where with a few minute's warning the roof can be closed.

Buildings are now designed so that tenants can directly monitor and control many operational features through their own PCs. The Tango housing complex in the Western Harbor redevelopment district in Malmo, Sweden is a case in point. Part of the international building exhibit there (Bo01 Expo) that took sustainability as the primary theme and goal, the Tango complex utilizes a variety of IT concepts. Tenants can directly control and program lighting, heating and ventilation, and security functions, through the complex's private service portal. Each apartment includes an "IT cupboard" where among other things a transmitter providing wireless service is found. Tenants can change the electronic text outside their apartment doors and on central electronic message board. A number of building services are accessible through the web portal: tenants can reserve the building's communal room and guest apartment in this way, can access information from the weather station, and monitor power consumption, among other functions. Tango has a number of other impressive green features and is part of a broader strategy for ecological regeneration for this area. Tango itself has a green roof and solar panels that feed hot water to an impressive district heating grid for the area.

The prospect of a continued reduction in the cost of computer chips suggests new ways in which a building and its contents can be intelligent. Electronic tags can be useful in many ways, for example, for tracking building materials on a building site. Tags that monitor occupants – their number and location – can add important safety dimensions to a building. Fire departments might now have critical information about where and how many occupants there are in a structure providing clues and information that can guide evacuation and other on-the-stop emergency decision making.

Singapore's Capital Tower, a 52-story downtown office building, allows occupants to make a variety of adjustments such as the temperature of their offices or reserving an executive elevator, directly through their PCs and a high-speed network (100 Mbps) (Horwitz 2002). Workers in the building can do many things in this way, including accessing security cameras, even reserving a parking space through an automated carpark system. As Horwitz (2002: 38) explains:

> For example, when a tenant pulls up to work in the morning, he or she enters the garage and a welcome display informs them as to how many, if any spaces are currently open, and on which levels. If the tenant is an executive, the vehicle identification system will open the appropriate gates to lead the tenant to his or her reserved parking space. For other employees, the system directs individuals to the closest available parking space.

One step beyond is the incorporation of small circuit boards into the walls and various key spaces in a building or home. Researchers at MIT's Center for Bits and Atoms have been developing such a technology under the label "Internet O" (Daviss 2003). The key idea is to essentially turn a building into a computer, and to create Internet-style connectivity between every room and system in the structure. Whether this is the correct system to do the job or not, there is no question that the future of intelligent buildings will be about re-conceptualizing them as highly web-connected, flexible living and working spaces.

There has been considerable interest in *smart homes* in recent years and substantial product development and research here (e.g. Gann *et al.* 1999). Smart homes are certainly much like smart or intelligent buildings more generally and have tremendous potential for reducing energy and resource consumption in very similar ways. Some homes have computer-guided geothermal heating and cooling systems. The Innovation Center for the Intelligent Home, or *inHaus*, based in Duisberg, Germany, opened in 2001 and is testing a variety of these smart and green features. Main elements include the use of sensors to control lighting and to open and close windows. Reduced energy consumption in these homes has been a special emphasis and goal in this work:

> Sensors in the rooms measure the temperature, humidity and air quality, and automatically open and close the motor-driven windows. If a thunderstorm approaches, the house automatically makes sure that everything is weatherproof. When the occupants leave the house, the heating switches itself down to economy level. Electricity costs are reduced by active load regulation, which for instance starts the washing machine during off-peak hours at night. The occupants of the house have constant feedback on their energy consumption, enabling them to correct their usage habits. Even the inHaus garden – its water sprinklers and lighting system – is integrated in the resource management program (Miller 2001: 31).

The *inHaus* project is also experimenting with effectively connecting the private automobile with the house. An intelligent car might provide the ability to communicate with and control many functions in the house on the road and through the auto. Conversely, smart car features might include automatic seat and mirror adjustments, the downloading of desired music from the home entertainment center, among other clever potential ideas.

Many of the smart home features seem especially suited to hurried and harried lifestyles. Prototype smart homes in the Swedish community of Varmda, near Stockholm, have so-called Unattended Delivery Units (UDUs), which allow delivery of groceries and other Internet-ordered goods when the homeowner is not there to receive them. Access to the box is through a keypad combination. Smart homes would allow their occupants to remotely lock and unlock doors. In some smart home prototypes front doors can be unlocked with fingerprints. One's computer might automatically check the traffic in the morning as the occupants arise, perhaps mapping out an alternative route to work to avoid the worst traffic. Voice controlled faucets, bathtubs that fill up to the appropriate level and no further and that set the water to a pre-determined temperature are some potential features of the smart home of the future. Self-propelled lawn mowers may one day "roll out of their storage sheds at preset intervals and mow the grass by following sensors embedded in the ground" (Williams 2001: A1).

Among the emerging ideas about what a smart home is, are notions of connecting all of the appliances and electrical systems in the home so that they can communicate with one another. The refrigerator in one's home could become the "nerve center" for the connected network. This is the idea behind the Screenfridge, a commercially available Internet-fridge manufactured by Electrolux. A flat screen is designed into the door of this fridge, and becomes the "platform and connection center for the networked features of the IT house" (Irish Times 2002). Such Internet fridges also have the potential to take advantage of electronic labels on the products stored in them – continuously taking stock of what might need to be replenished as well as potentially sending alerts when product expiration dates approach or pass.

Whirlpool's Polara oven, is another example of a smart Internet-enabled appliance. This unique appliance is equipped with a refrigeration compartment and is Internet-ready – simply, the user can remotely instruct the oven to shift from refrigeration to cooking, with the meal ready when the occupants arrive home at the end of the day (Lorek 2003). These instructions can be sent to the appliance with the resident's Internet enabled cell phone.

IT has the potential to address a number of current obstacles that make certain aspects of green or sustainable living difficult. IT can aid independent mobility and lifestyles for older and younger members of society, for example. As the American population continues to age (indeed the global population), new challenges exist to ensuring safe, dignified living conditions for older residents. IT and intelligent buildings and neighborhoods can make it easier and more feasible, for example, for older residents to "age in place." Examples include elements in the design that allow easy communication between doctors and other health care workers, and that can detect situations where medical assistance may be needed. In an IT home, sensors in the bed and motion sensors in each room can detect the need for lighting and thus avert nighttime falls. Similarly, sensors might detect when a stove burner has been left on – turning it off or alerting the resident or caregiver that there is a potential danger. Panasonic has even been developing a "healthy bathroom system," which could "automatically measure your weight, body fat ratio, and urinary sugar whenever you use the toilet. The health data is stored in the home server for future reference, and may be sent through the network to the diet service or other health consultants for analysis and advice" (Smart News Online 2001).

Particularly promising is the idea of clothing-integrated sensors connected to a centralized monitoring system that would detect if emergency circumstances or health problems arise in older residents. Smart or intelligent appliances in the home hold the promise of making independent living much easier and safer. A truly smart or intelligent home would monitor and learn from the movements and behavioral patterns of its inhabitants, for instance knowing the extent to which the heat is usually turned up in the morning or when lights are turned off in the evening. Once learned, these tasks would then be taken over automatically by the home's central computer.

Many of these IT systems and features, then, serve to reduce energy and resource consumption, and at the same time enhance living and working conditions for their occupants. Smart homes that switch lights on automatically for an elderly resident, thus reducing risk of a fall or injury and permitting more independent living, also result in energy conservation by reducing the need for lights to be on all night. Smart homes that adjust artificial lighting or temperature to maximize occupant comfort also typically reduce the structure's energy and heating needs. Smart homes are also energy-efficient and resource-efficient homes.

Intelligent design features and designing-in IT from the beginning should also make telecommuting and working from one's home more feasible. This is certainly also a green feature in the sense that it may reduce vehicle miles traveled, and in turn reduce energy consumed. Indeed, telecommuting and more flexible work arrangements are viewed by many as an important element in any successful response to urban traffic congestion in many metropolitan areas of the US Designing-in IT elements that permit videoconferencing, for instance, would be very helpful. Heavily wired homes and wireless hubs are increasingly a design feature sought by home buyers, allowing new levels of automation and control. Installing data ports into extra bedrooms that may later be used as home offices is a sensible approach. This indirectly leads to environmental improvements by reducing the amount of travel (assuming, however, that work trips are not substituted by other kinds of trips).

IT and Green Neighborhoods

Green design and intelligent design can also be applied at the level of the neighborhood or urban district; here, as well, there are many potential IT applications. There is increasing interest in the notion of energy-neutral or energy-balanced neighborhoods, where at least as much energy is produced as is consumed. Achieving energy-neutral or carbon-neutral neighborhoods entails a variety of green neighborhood strategies, including on-site renewable energy generation (e.g. rooftop photovoltaic panels, micro turbines, biogas harvesting from household wastes), energy efficiency, neighborhood food production, a transportation system that reduces reliance on the private automobile, and tree planting and nature restoration, among others. Many of these neighborhood strategies involve IT in one way or another.

Replacement of the highly energy-consuming, carbon-emitting private automobile with a different more sustainable neighborhood system is one area where IT could play an especially central role. Creative proposals have been made in recent years to create a system of on-demand IT-guided taxi-cabs and mini-buses. Visionary architect William McDonough had such a concept, though not yet realized, for the new community of Coffee Creek, Indiana, designed around green and sustainable principles. Here, residents would be equipped with GPS wristwatches that would permit the quick calling of a taxi, with but the press of a button on the watch (for a good review of many of McDonough's ideas, see McDonough and Braungart 2002). The growing interest in car-sharing might provide a similar opportunity to reduce reliance on privately owned autos. Here, a small neighborhood-based fleet of cars is shared by the community, permitting short trips and car usage when truly needed. The operation of these car-sharing systems is increasingly IT-based. Users reserve cars on-line, access the cars via smart cards, and receive a monthly bill for time used and distance traveled, with information collected from on-board computers in each car (for a discussion of car-sharing, see Beatley 2000).

Other forms of creative neighborhood mobility also might utilize IT. There is considerable interest in and innovation around the idea of public or neighborhood bikes in many cities in Europe and the US A system of collective bikes available free or for a small charge could represent a reasonable alternative to the car for many of our daily trips. In some cities, subscriber systems now provide residents with access to bikes through a smart card or electronic key. In Germany, the national train company Die Bahn has initiated a fascinating high-tech *smart bikes* program – making bikes available for quick rental and use at main train stations in several cities in that country. Known as *Call a Bike*, potential users call a central phone number to reserve a bike, paying with a credit card. Arriving at, say, the Frankfurt central station, users have access to one of the 25 or so distinctively shaped bikes found there. The bikes are each equipped with an electronic locking device, and renters access the bikes by punching in a unique code on the bike's built-in computer keypad. When finished, a user returns the bike to one of a number of drop-off points, and is automatically billed for the time the bike is used (at a rate of seven eurocents per minute). *Call a Bike* appears to work quite well and provides an important mobility option for residents and visitors alike.

Connecting communities and neighborhoods digitally could support and advance many important social and sustainability goals. Such a network might facilitate dissemination of information about new sustainable products and services, could help to coordinate community events and activities (from tree planting or stream cleanup to a community gardening workshop), and might help to promote sharing of goods and resources in a neighborhood. Moreover, it is conceivable that such a community portal, in its ability to provide a variety of real-time information, would help to support a variety of other sustainability ideas and technologies, for instance making public transit more attractive (e.g. with the ability to check whether trains are running on time).

One recent and notable example of a heavily "wired" neighborhood can be seen in Santana Row, a new mixed development in San Jose, California. Opened in the fall of 2002, it includes housing (1200 flats), retail commercial (120 shops), a cinema and a hotel, and illustrates how IT can be integrated into a new neighborhood or urban district. The "technology backbone" of this neighborhood is the plug-and-play network found throughout the community. High speed fiber-based Ethernet connects are made between the homes, shops and community. Santana Row is described as an "intelligent community for residents and retailers," where "high-speed Internet and digital cable television were as basic as public utilities like water and electricity" (Cisco Systems 2002: 3).

One of the important elements is an "on-line community portal" that is intended to provide residents and business tenants with a variety of community information. Described as a "virtual meeting place for community information and interaction," it provides community news, an events calendar, and the prospect of a single place where useful information about the community can be found and where residents, guests and businesses can easily communicate with each other (Cisco Systems 2002: 5).

One of the most interesting features of Santana Row is its provision of wireless Internet access throughout all of the public spaces in this community:

> Santana Row placed the access point antennae to provide coverage from many of the community's park areas, as well as several public areas in buildings, including outdoor dining areas in restaurants, coffee shops, cafes, and hotel lobbies. The antennae were fully enclosed within the buildings to maintain aesthetics, so the wireless network remains invisible. And, because the access points were positioned to provide overlapping coverage, users can wander freely throughout the community, checking mail or surfing the web – all without losing their network connection (Cisco Systems 2002: 4).

The Hotel Valencia, a 213-room hotel in Santana Row, incorporates IT in several creative ways. These include the use of a remote wireless check-in system that allows guests to completely register curbside, sending them directly to their rooms.

GPS and tracking technologies may in some important ways allow Americans to live healthier, more secure, more sustainable lives. As Americans continue to get heavier – more than 60 percent of Americans are now considered by the Centers for Disease Control to be overweight – our sedentary lifestyles are beginning to catch up with us. The paucity of walking, bicycling and other outdoor physical activity is shocking, and the resulting over-dependence on cars represents a major environmental sustainability challenge. GPS and tracking technologies might help in several ways. For instance, concern about the safety of children prevents many from walking to school. While just 30 years ago, two-thirds of children walked or rode their bike to school, that number is today a dismal 13 percent. Being able to locate and track a child's movement might offer the possibilities of overcoming such safety and security concerns. Several companies now offer GPS bracelets or backpack systems that permit the tracking of children. Wherify Wireless, for instance, offers a personal locator for about US$400 and a US$25 monthly fee. Worn on the wrist, it resembles a "digital watch pumped up on steroids" (Wade 2003: E6) and permits parents to monitor the location of a child even via the Internet.

Obstacles on the Road to Intelligent Buildings and Neighborhoods

Given the recent advances in IT and its tremendous potential benefits – economic, social, environmental – it is perhaps surprising that full application of these technologies remains limited. There are unfortunately a number of obstacles and impediments faced. These include a tendency to focus on the short-term building and design costs, rather than viewing the construction of homes, office buildings and neighborhoods as "long-term investments" (Paumgartten 2002). A new "whole building perspective" takes a different tack, one that looks at the entire life cycle of the building.

One approach to overcoming this obstacle is for state and local governments to provide financial incentives that serve to reduce these initial upfront costs. Arlington County, Virginia, now provides a density bonus for green buildings under its zoning code. Public utilities such as Austin Energy are increasingly providing rebates and purchase credits for green technologies, certainly including IT elements. The State of New York now provides a tax credit for green building. These are important steps in leveling the playing field and giving a financial boost and encouragement to green design.

Greater public education is certainly a primary element in overcoming such obstacles. This must include education of a number of constituent groups – builders and developers, architects and design professionals, housing consumers, among others. Greater attention to green design concepts and principles in schools of architecture, including the application of IT, would be beneficial. Educating housing consumers and making the energy and resource implications of their choices clearer would certainly also help. This education can happen in many ways, including through model green homes, such as the Green Homes for Chicago initiative. Here the City funded a national design competition for green homes and, once selected, has now funded the actual construction of the five designs chosen. Chicagoans will have the chance to see what a green home looks like, and builders and architects have been encouraged to see how green principles might shape their future projects.

Concerns about privacy undoubtedly remain a serious obstacle to incorporating a number of the IT ideas discussed here. Fears about the collection of personal data and over-surveillance of individual behaviors and lifestyles are a concern for many. The same IT system that permits monitoring of the health and safety of an elder resident, and perhaps permits that individual to live independently, also involves a potentially significant degree of infringement of personal privacy. These concerns suggest, among other things, the importance of special safeguards on the collection and dissemination of personal information (e.g. digital data about one's library book selections, or about one's consumption or other lifestyle patterns).

In advancing green design and intelligent buildings and their IT dimensions, one must carefully consider and understand the broader societal and cultural impacts of these ideas. Creating living arrangements that promote delivery of goods to refrigerated lock boxes that permit individuals to work from their homes and that allow an automated lawn mower to take care of the yard work, to name a few of the clever ideas here, may not necessarily be good ideas from the perspective of building social capital in a community and strengthening ties between people. We must always be careful about such unintended consequences and to place green and intelligent design in the context of these broader social goals and commitments.

References

Beatley, T. (2000) *Green Urbanism: Learning from European Cities*, Washington, D.C.: Island Press.

Beatley, T. (forthcoming) *Native to Nowhere*, Washington, D.C.: Island Press.

Building Design (2003) "Business and IT: Smarter than the Average," *Building Design*, March 21, p. 20.

Cisco Systems (2002) "Santana Row: Building an advanced networking community," Cisco Systems and HP white paper, San Jose, CA: Cisco Systems.

City of New York (1999) *High Performance Building Guidelines*, Department of Design and Construction, April.

City of New York (2002) *Implementing the High Performance Building Guidelines*, Department of Design and Construction.

Daviss, B. (2003) "The Net Comes Home: Strip the internet down to its essentials and you have the perfect backbone for an intelligent building; Bennett Daviss went to MIT for a demonstration," *New Scientist*, February 15, p. 26.

Entwicklungsgesellschaft (1998) *Mont-Cenis*, Herne, October.

Flame Retardancy News (2002) "Intelligent Buildings to Improve Fire Safety," *Flame Retardancy News*, Vol. 12, No. 1, January.

Gann, D., Barlow, J. and Venables, T. (1999) *Digital Futures: Making Homes Smarter*, Coventry, UK: Chartered Institute of Housing.

Glancey, J. (2003) "G2: Architecture: Space odyssey: From the outside its sleek as a rocket but is the Swiss Re tower really as sensational as it looks?" *The Guardian*, December 8, p. 12.

Hathway, W.E., Hargreaves, J.A., Thompson, G.W. and Novitsky, D. (1992) "A Study into the Effects of Light on Children of Elementary School Age: A Case of Daylight Robbery," prepared for the Alberta Education Department, Online. Available at: www.naturallighting.com (accessed January 12, 2004).

Heerwagen, J. (2002) "Sustainable design can be an asset to the bottom line," *Environmental Design and Construction*, Vol. 5, No. 4, pp. 35–40.

Heschong Mahone Group (1999) *Daylighting in Schools: An Investigation into the Relationship Between Daylighting and Human Performance*, Submitted to PG&E, August.

Horwitz, B. (2002) "Singapore's Savvy New Tower," Reed Business Information Service, September 1, p. 38.

Irish Times (2002) "Home Smart Home," *The Irish Times*, January 12, p. 51.

Kats, G. (2003) "The Costs and Financial Benefits of Green Buildings," A Report to California's Sustainable Building Task Force, October.

Lorek, L.A. (2003) "Intelligent Homes in the Future," *San Antonio Express-News*, August 6, Metro, p. 1-C.

McDonough, W. and Braungart, M. (2002) *Cradle to Cradle: Remaking the Way We Make Things*, New York: North Point Press.

Miller, F. (2001) "Wired and Smart: From the Fridge to the Bathtub," *Fraunhofer Magazine*, Vol. 2, pp. 30–2.

Paumgarttern, P. (2002) "Perspectives on Sustainable Design," *Environmental Design and Construction*, November.

Rocky Mountain Institute (RMI) (1998) *Green Development: Integrating Ecology and Real Estate*, New York: John Wiley and Sons.

Romm, J. and Browning, W. (1998) *Greening the Building and Greening the Bottom Line: Increasing Productivity Through Energy-Efficient Design*, Snowmass, CO: Rocky Mountain Institute.

Smart News Online (2001) "Panasonic's Future Home," *Smart News Online*, Online. Available at: www.smartnewsonline.com (accessed January 16, 2004).

US Green Building Council (undated) "LEED Fact Sheet," Washington, D.C.: USGBC, Online. Available at: www.usgbc.org (accessed January 12, 2004).

Wade, W. (2003) "Keeping Tabs: A Two-Way Street, Finding your children has gotten easier, but then, so has finding you," *New York Times*, January 16, p. E1.

Watson, S. (2002) "Turning to Technology," *Buffalo News*, January 18, p. A1.

Wigginton, M. and Harris, J. (2002) *Intelligent Skins*, Oxford: Butterworth-Heinemann.

Williams, C. (2001) "Meet Jorgen Jetson – Swedes Take Lead in 'Smart' Living; Lifestyle: Nation is embracing high-tech devices that are expected to sweep the developed world," *Los Angeles Times*, April 6, p. A-1.

CHAPTER TWELVE

Infrastructure in the Age of Earth Systems Engineering and Management (ESEM)

Braden Allenby

Introduction

The continuing integration of information technologies (IT) into physical infrastructure raises a number of interesting issues at many scales. For the most part, as demonstrated by the other chapters in this volume, these issues can be addressed through existing institutions and disciplines. Increasingly, however, these systems are also beginning to integrate natural, built, and social science systems at higher levels, resulting in emergent characteristics and challenges that are broader than existing practices can comfortably manage. This has led to the evolution of the concept of earth systems engineering and management (ESEM), which has the potential to provide a powerful augmentation to current approaches to infrastructure design and operation. Here, two integrative processes will be of interest: first, the integration of information processing and communication capabilities in all forms of infrastructure at many different scales; and second, the integration of human and natural systems, with special emphasis here on that trend as it regards infrastructure. This chapter will begin with a brief introduction to earth systems engineering and management, then continue with an exploration of how this approach can begin to inform the evolution of "intelligent infrastructure."

Earth Systems Engineering and Management

It is increasingly apparent that a principal result of the European Enlightenment, the Industrial Revolution, and the accompanying changes in human demographics, cultures, technologies, and economic systems, has been the evolution of an

Earth in which the dynamics of major natural systems are increasingly dominated by human activity (Allenby 2002). Reasonable stability of these integrated human/natural systems requires the development of an ability to responsibly, rationally, and ethically design and manage them. This capability is the objective of "earth systems engineering and management," defined as "the capability to rationally engineer and manage human technology systems and related elements of natural systems in such a way as to provide the requisite functionality while facilitating the active management of strongly coupled natural systems" (Allenby 2002: 8). Both the understanding and practice of ESEM is, however, fairly primitive at this point, with significant weaknesses in the current science and technology base, institutional and governance structures, and relevant intellectual disciplines and methodologies. Indeed, in many cases there is a reluctance to either perceive the current realities of an anthropogenic world – what *Nature* (2003) calls the "anthropocene" – or to accept the need for new thinking. And this is not necessarily irrational, for there are constant claims that conditions are "new" or "unique" for some reason, most of which are poorly supported at best. Thus, the starting point of any new approach to an area such as this that has already been well studied in many ways is to return to basics: What has changed contextually that enables, or even requires, new thinking? This is particularly appropriate when the new approach differs not just in degree, but perhaps even in kind, from what has gone before it. In beginning this process, it is useful to return to basics, and perhaps a certain intellectual humility. As Heidegger reminds us ([1952–1962] 1977: 49, 136): "So long as we do not, through thinking, experience what is, we can never belong to what will be. The flight into tradition, out of a combination of humility and presumption, can bring about nothing in itself other than self deception and blindness in relation to the historical moment."

Illustrative examples of ESEM are legion. Consider, for example, the issue of global climate change, arising in large part from current transportation and energy technologies and infrastructure. The problem is frequently stated in terms of reducing energy use, an ideological approach that is both politically and practically difficult to implement. But the real challenge is both more daunting, and more fundamental: the very existence of a climate change debate recognizes that human impacts on both the climate system, and the carbon cycle, are so substantial that integrated management is necessary. Similarly, consider the case of the Florida Everglades. The current biological structure of the Everglades arises in large part from changes in water quality and quantity that, in turn, are based on both agricultural and demographic trends (the US sugar industry, and urbanization of Florida coastal areas, respectively). Thus, while the restoration effort involves significant civil and hydrologic engineering, this in turn implies design objectives and constraints involving economics, trade, politics, and demographics. In short, one cannot engineer the Everglades without implicitly having to design much of Florida – and not just along civil engineering lines, but in cultural, political and economic dimensions as well. In both cases, major infrastructure elements are involved – transport and energy in the global climate change area, and civil and hydrologic engineering in the Everglades. But in both cases, implicit in the infrastructure design and operation are far more complex and profound engineering challenges.

Integrated Infrastructure and Complexity in the Anthropogenic World

In these as in similar cases (designing urban systems, for example), a principal theme is greatly increased complexity, and a principal question is what fundamental issues in conceptualization and understanding are posed by civil infrastructure and research involving infrastructure. To begin with, there are at least three kinds of complexity that one must anticipate, although all may not be present in any particular project. The first, and the direct subject of this volume, is the integration of IT and infrastructure engineering, and its effects on the systems involved. In "dumb" infrastructure, components seldom relate to each other except mechanically and through the supervision of human institutions. But as "smart" infrastructure evolves – that is, as systems grow to integrate sensor, communication, and information processing into the infrastructure itself – the very nature of the infrastructure changes. For one thing, previously disparate components couple through information systems into higher level systems, and begin displaying emergent behaviors that are difficult, if not impossible, to predict based on component performance. An interesting example is the automobile (Allenby 1999). It used to be an artifact, connected to its external environment and supporting infrastructure solely by the whims and perceptual apparatus of the driver. Now, of course, the automobile is an information artifact, with sensors that tie its energy systems to ambient atmospheric conditions, and mapping systems that tie it, through Global Positioning Systems, to transportation infrastructures and even, through real time traffic information, to current conditions of that infrastructure (such as traffic jams). This integration of artifact with infrastructure, through greatly enhanced IT systems, has a number of effects. Most notably, it enables seeking design and operating efficiency at much higher levels of the system where emergent behaviors such as traffic jams can be addressed directly (by rerouting traffic in real-time, for example). Or consider the Internet. It is clearly a purely human product: routers, fiber optics, and computers are all artifacts of human design and intentionality. But once they are assembled into an operation, no person knows what kind of behavior will result – indeed, the system becomes self-organizing and evolves in ways that the engineers of the various pieces cannot predict. As Barabasi (2002: 145, 149–50) points out, "[t]hough human made, the Internet is not centrally designed. Structurally, the Internet is closer to an ecosystem than to a Swiss watch ... [w]hat neither computer scientists nor biologists know is how the large-scale structure emerges once we put the pieces together."

To add to this growing systems complexity, once infrastructure is infused with information systems (sensors, information networks, information processing capability), it can increasingly be designed to respond to perturbations without human intervention. Thus, for example, communications networks increasingly are being designed to be "self-healing" – to use the intelligence and inherent resiliency of the network structure to automatically maintain functionality in the face of disruption.

The second form of increased complexity arises from the evolution of the external policy and social context within which infrastructure design, construction, and operation takes place. Not just governments at all scales, but communities and, increasingly, non-governmental organizations (NGOs) of all kinds, especially human rights activists and environmentalists, are routinely involved in infrastructure processes – and usually with mutually exclusive positions, at least initially (Mathews 1997). Engineers are generally not taught to manage stakeholders; given how involved especially civil engineers get in contentious projects, from restructuring the drainage of the Everglades to The Big Dig in Boston, this is unfortunate now, and will become more so. "Smart" infrastructure may operate at more integrated scales with greater information processing power, but the ideologies and teleologies that drive it remain profoundly human.

The third form of increased complexity arises from the impact of new kinds of information – specifically, information on environmental and social dimensions – on the design, building, and operation of projects. While most notable, perhaps, in obvious cases such as building gas pipelines in Latin America or airport facilities in crowded urban areas, it is increasingly seldom that infrastructure of any kind is unaffected by new arguments and concerns in non-engineering areas, usually introduced by parties not directly involved in the projects. Thus, not only "technical" complexity, but political and cultural complexity, increases as infrastructure becomes more intelligent, and, reflecting new systems dynamics and increased scale, integrates human and natural systems in new ways as well.

In general, as Heidegger suggests, we are failing to perceive precisely how complex and challenging the world we live in right now actually is, and how it is evolving to become even more so. Our ignorance in this area, trapped by preconceptions and outdated ideologies, is both profound, and invisible to most people. Consider, for example, the evolution of an increasingly wired world. Combine sensor systems, nano-technology, the evolution of an environmentally sensitive culture, and enhanced IT, and what you easily come up with is a world which is constantly monitored on virtually all scales and along all relevant parameters: CO_2, nitrogen, hydrologic and climatic systems; and economic, cultural and technological systems. In other words, IT networks change natural systems, and the world itself, into a kind of infrastructure. Now ask yourself a couple of basic questions. One obvious question is: When you wire a world like that, what do you do to our perception of the world? What do you do to culture? If you know that even when you are in the middle of Yellowstone, you are surrounded by sensors that are feeding into a NASA computer tracking in real time the environmental parameters of where you are, and that you yourself are constantly tracked, and available through your IT system, as well, do you get the same sense of what Yellowstone is? Does "nature" mean the same thing to you when there is no "wilderness" left in the world? Certainly for those that hold "nature" as "sacred" there may be an element of blasphemy about such a world (McKibben 1989; Allenby 2002). And how will the building of such an infrastructure change what it means to be human – indeed, how can we learn to think about infrastructure as a mechanism for altering human culture, even morality?[1]

Heidegger cautions us not to be trapped by perceptions that reflect past verities, not present realities. Accordingly, let us consider the implications of such a sensored world from another perspective, and ask the provocative question: as humans continue to evolve, is there any element of the present world that does not become, implicitly if not explicitly, a product of human design and choice? In short, is it not an anthropogenic world? Perform an easy thought experiment. Think of an alien looking at our Earth for the first time from its spaceship perspective. What is the most immediate and predominant impression it would get of the world? Perhaps most fundamentally, this view of the Earth reveals that it is in large part a product of design: the emissions spectrum reflects lights, radio and television, communications networks, and heat emissions from cities; fundamental chemical and physical cycles are impacted by human activity; the biosphere reflects centuries of agriculture, settlement, land use, and, increasingly, engineering at the genomic level. Are there areas of wilderness? Sure. But, like the Florida Everglades, the Alps, or Antarctica, they exist at our suffrage, and increasingly reflect deliberate choice. The world is not a human artifact, but it is a planet that increasingly reflects in the dynamics of most fundamental systems the activities of one species – ours (Turner *et al.* 1990; Vitousek *et al.* 1997; Allenby 2002).

Under such circumstances, what is infrastructure? To begin with, it is apparent that the kinds of things we traditionally think about as infrastructure – transportation networks, water and sewage systems – clearly remain important. And much design and continuing integration of IT with their functions can occur within the current intellectual and disciplinary boundaries by which they are defined. But we can also begin to identify emergent behavior as these built systems integrate in ever more complex ways with each other, the natural environment, and other human systems, at higher scales. For one thing, we begin to find ever greater inner-connection of human and natural systems – the carbon cycle, the sulfur cycle, certainly the hydrologic cycle – with the natural systems more and more becoming elements of the human infrastructure. We integrate them into our economic activity, we integrate them into our planning activity, we integrate them into our political activity. Thus, for example, genetic engineering is integrating previously independent biological systems – in this case, biological structures at the genomic level – into food production and, through politics, food distribution systems. The global climate change negotiations are, in part, a mechanism by which portions of the carbon cycle are brought into the human economic system and made human. And returning to the sensored world scenario, what we see is a process that takes a lot of the information that previously existed in non-human systems – the climatic and atmospheric systems, biological systems at all scales, elemental cycles and the like – and makes it part of the human infosphere. At this point, a profound integration of IT, natural systems and "natural infrastructure," and human cultural and technological systems begins to develop, and natural systems become subject to human manipulation. We still have traditional infrastructure, perhaps augmented by IT in new and complex ways – but now we also have infrastructures developing at higher levels of the system, in ways that we need to learn to understand.

Information Infrastructure: a Case Study

Given this increasing complexity, how should one think about infrastructure? Using information infrastructure as an example, it is possible to develop a generalized approach that identifies a few easily definable layers (at least conceptually). Developing such a schematic – and, obviously, this is only one approach – is important, at least in part because it helps reduce the tendency to confuse the issues, concepts, and methodologies appropriate to different layers. As one example, early in the development of the Design for Environment module of the DFX design methodology ("Design for X") (AEA 1992), a design team at a large electronics manufacturer was asked to develop a "sustainable telephone." No one, including the people that were asking for the product, had the faintest idea what that meant, in large part because "sustainability," whatever it is, is a property of global systems, not individual telephones. When, however, the challenge was broken down into heuristics – less energy consumption, less use of toxics, less use of packaging – it was easily integrated into the design process.

The obvious beginning point for such a schematic is the individual manufactured product: after all, even in the case of a service industry, such as telecommunications, any service is based at some point on physical artifacts, whether they be buildings, manufactured complex products, substances used up in providing the service, such as fuel, or the like. Here, several important stages involving artifacts can be identified. Thus, the first stage is the design, and the second the manufacture of the artifact. These two stages are coupled and, in many cases, will form an infrastructure on their own, with manufacturing processes drawing heavily on materials and components from around the world, and thus relying on associated transport, energy, and communications systems. This makes an important general point about infrastructures in the anthropogenic world: they are increasingly coupled together in a complex web of global trade in materials, energy, components, subassemblies and assemblies, which is in turn heavily coupled to complicated financial networks. Linking these systems is an increasingly complex overlay of information structures, which not only enable the overall system to evolve, but are a primary source of complexity and systems dynamics in their own right. Thus, the first and most simple level of any infrastructure – the individual constituent artifacts – already implies a network of supporting infrastructures, from transport to finance to IT.

There are an increasing number of complexities associated with even these preliminary stages, of course. These may be fairly technical – for example, making the choice between lead-based solders versus those based on bismuth or indium requires that the electronics designer understand the availability and ore characteristics of each option. But in a world increasingly concerned with environmental and social dimensions of industrial behavior, and with activist communities increasingly able to draw on information about such impacts from around the world, this relatively technical approach to supply infrastructure is becoming inadequate. Thus, knowing that bismuth is primarily a by-product of lead mining, and occurs in relatively low concentrations, indicates that such a choice may increase the local environmental impacts of mining and energy consumption required to produce the purified metal (Allenby 1992). Similar questions arise when the process designer chooses among potential cleaning system alternatives – say, between chlorinated solvent and aqueous solvent systems. The designer will understand as part of the job which options perform as technically required – but will also need information such as how much energy is embedded in the materials chosen, to understand the broader social and environmental impacts of the choice. And, increasingly, understanding such dimensions is important to designing a robust supply and manufacturing infrastructure.

Individual information artifacts by themselves are relatively useless, of course. They become useful as they are attached to, or support, physical networks that enable communications. This is the level that is classically thought of as "information infrastructure," although it is already apparent that any infrastructure is in fact a network of coupled systems, each of which has its own characteristics, strengths and weaknesses, and each of which must be considered in evaluating the robustness, and environmental and social preference of the infrastructure as a whole. The type of engineering, and the knowledge base required, shifts at every level, however. At the network level, for example, the network operator, whose expertise involves network design and optimization, not artifact design, will not, except in rare instances, attempt to impose specific product design material requirements. That is because the tradeoffs involved in, for example, selection of each material will be known to the artifact designer, but are not within the knowledge of the network operator. Moreover, optimization of the economic, technical, environmental and social performance of the network as a whole – for energy consumption per unit of information transmitted, for example – will be a network system function, not a specific artifact function.

At this point, another foundational principle of dealing with complex systems of intelligent infrastructures is worth noting: the system of interest is defined by the question asked. Thus, for example, if I am thinking of New York City, and I ask how it manages political questions, the physical boundaries implied by the question are primarily the usual jurisdictional ones: the five boroughs, for example. If, however, I am a civil engineer and I ask about the water system of New York City, the system boundaries implied by the question include about two-thirds of the State, for that is the watershed area feeding into New York City's water supply – an area that, for at least 100 years, has had its settlement patterns, land use patterns, and built infrastructure in part determined by New York City's need for water. If, perhaps, I ask about management of waste in New York City, I make Pennsylvania and Virginia, and connecting highway routes, parts of the system, for New York City waste is shipped to those states for disposal. In other words, what infrastructure I consider relevant is defined by the question I pose to the undifferentiated complex system in front of me.

Traditional assessments might hold that the physical network level is the highest level of interest in evaluating IT infrastructure, but arguably the most important impacts, and design implications, of an information infrastructure will be missed if one simply stops there. The services supported by the physical infrastructure constitute yet another level of infrastructure and, while usually much more difficult to evaluate, are perhaps even more fundamental in their economic, social and environmental impacts. For example, many telecommunications services provide functionality that in turn significantly impacts demand for materials. In the case of corporate intranets, for example, the publication of much corporate information in cyberspace rather than on paper significantly reduces paper demand in many companies (as just one illustration, AT&T saves tens of millions of sheets annually by publishing its corporate human resources, environment and safety, and other internal documents on its web site, rather than on paper in large binders to be provided to each manager). Capturing the effects of such services on material stocks and flows, and identifying opportunities for future technologies and efforts, is an important element in understanding the value, implications, and design dimensions of infrastructure at this level.

Evaluating the services level is made more complex because frequently the services provided by telecommunications companies enable the evolution of cultural patterns that, in their turn, affect other infrastructures, and material stocks and flows in the longer term. One example is telework, a cultural practice enabled by the development of virtual private networks that ensure remote locations can easily and securely connect into corporate information systems. Telework in turn impacts patterns of consumption of gasoline and, more gradually, will undoubtedly affect the way houses are designed and built (more space for home offices, thus changing building material demand), as well as the spatial pattern of built environments over time, thus affecting infrastructure design and material requirements. More subtly, it enables continued participation of seniors in the job market, with impacts ranging from increased tourism and travel in that age group, to reduction of the economic challenge of increasing pension liabilities as the baby boomer generation retires (Allenby and Roitz 2003). Tracking such impacts is very difficult, requiring both social science and physical science expertise, as well as knowledge of underlying material consumption patterns. To date, research on such highly multidisciplinary issues is sporadic and badly underfunded, and most engineers are not trained to evaluate such broad effects. It bears repeating, however, that the purpose of this approach is not to replace traditional infrastructure engineering skills; they are still required at the project level. Rather, it is to augment them to reflect the increasingly complex technological, social, cultural and economic context within which infrastructure conceptualization, design, engineering and operation are now occurring.

But even the services level may understate the complexity of the integrated infrastructure systems that arise as a characteristic signature of the anthropogenic

Earth. Following Heidegger's admonition not to be blinded by our existing beliefs as to what is actually current reality, let us consider for example the implied relationship between human systems, including infrastructure, and "nature." Clearly, for most of human history the relationship between the human and the greater world is that of the contained in the container. In the vernacular, the human world would be understood as subordinate, and completely within, the "natural" world. Even from the beginning, of course, certain elements of nature are made part of the human – even primitive agriculture does that. But with growing urbanization, and consequent industrialization of the world, the implicit mental model becomes one where the human sphere is relatively divorced from the natural sphere. This, in fact, is what enables the romanticizing of "nature" into a non-human, sacred realm, a process that gained ascendancy in the Romantic tradition of the European Enlightenment (Abrams 1971). Before this mental model became dominant, the opposite of "natural" was "supernatural" – that is, ghosts, goblins, and other creatures that didn't exist. With the dominance of this mental model, however, the opposite of "natural" becomes the human, which is split apart from the "natural." Thus, the opposite of a "natural" food is a genetically engineered one; the opposite of a "natural" material is a "man-made" one. Indeed, this dichotomy is a very powerful element of the environmental movement – and one reason that urban systems, and urban infrastructure, tend to be reflexively opposed by environmentalists (Cronon 1996).

Now, however, that mental model is obsolete. Rather than "nature" and "human" being completely separate, the "natural" is being increasingly integrated into human systems of all sorts, from urban, to economic, to technological and, as the climate change negotiation process demonstrates, the political. Not only are the two kinds of systems increasingly integrated, but the human systems are becoming more and more dominant. It is not that humans control the world in some explicit sense, because we have neither the knowledge nor the skill to do so. But the human impact becomes a significant element of the dynamics of most fundamental natural systems. It is in this sense that infrastructure is gaining increased importance: it is not just a complex engineering and design challenge in itself, but by doing so we are also designing, albeit perhaps inadvertently and unintentionally, coupled natural systems as well. Thus, what engineers do to "recreate" the Everglades in fact is a new design process, with the design objectives and constraints now including on-going management of many aspects of the Florida region as a whole previously considered disparate. What engineers may do to create new energy infrastructures, or active carbon sequestration technologies, becomes part of a carbon cycle management process.

This, then, is the ultimate level of complexity, where infrastructure becomes integrated through complex sets of information structures (including not just formal IT systems but, e.g. information systems such as the market economy), both across differing forms of infrastructure, and with "natural" systems as well. And, as a moment's reflection on IT infrastructure demonstrates, the potential issues and impacts at this level are profound, if not well understood and, indeed, not even perceived.

For at this level, the complexity and emergent behaviors are not simply in the technological realm. It is not just that we can't map the underlying physical structure and dynamics of the Internet. It is that the Internet, like any foundational technology, will have profound impacts on human society and culture. Thus, for example many of the terms that characterize the environmentalist discourse, and that are taken as self-evidently "real" – "sustainability," "nature," "wilderness," or "environment," for example – are, in fact, highly contingent cultural constructs. That is, they reflect not eternal verities that are "out there" as much as they do particular ways of thinking about the world that reflect a particular point in cultural and historical evolution. As Merchant (1995: 153) observes, "Nature, wilderness and civilization are socially constructed concepts that change over time and serve as stage settings in the progressive narrative." Perhaps the most obvious example is the concept of "sustainable development," which is clearly an explicit cultural construct: after all, it was invented and popularized in a specific book, *Our Common Future* (WCED 1987). Over the past 15 years, however, the cultural construct of "sustainable development" and the looser term of "sustainability" have become a principal goal of environmentalism: what began as a cultural construct now defines the desired endpoint for all human activity – the teleology of sustainability. In the process, the contingency of the term – although quite explicit – has vanished. Why bring this point up? Because the infrastructure of the Internet bids fair to increase the rate of change in cultural constructs by accelerating the rate of information flow and evolution through society and the economy – an effect that is hard to discuss rigorously, and whose outcome cannot be predicted with certainty.

We can trace a similar evolution in the cultural construct of "wilderness." In the Bible, as in most pre-Romantic literature, "wilderness," as opposed to gardens or agricultural areas, was evil. Indeed, for many early Enlightenment Romantics the source of "wilderness" and the wild – viewed at the time as a flaw in Creation – was the Biblical Fall (Abrams 1971: 99):

> . . . a perfect, wise God had originally created a perfectly smooth, orderly, useful, and beautiful world. Mountains and other wild, waste places were the product not of divine benevolence but of human depravity, for they had been wreaked by the wrath of a just God at the original fall of man in Eden, or alternatively (in some commentors, additionally), they had been effected by the devastating flood with which he punished the all-but-universal corruption at the time of Noah.

Not surprisingly, then, the first European settlers arriving in North America carried with them this cultural construct of "wilderness;" in fact, the pioneer was heroic precisely because she or he tamed a wild, empty wilderness, and made a fruitful agrarian environment of it (Sagoff 1996). The common cultural view was expressed, for example, in John Quincy Adam's 1846 appeal to Americans to settle Oregon, "to make the wilderness blossom as the rose, to establish laws, to increase, multiply, and subdue the earth, which we are commanded to do by the first behest of the God Almighty" (Quoted in Merchant 1995: 143). Ralph Waldo Emerson was even more vigorous:

> This great savage country should be furrowed by the plough, and combed by the harrow; these rough Alleganies should know their master; these foaming torrents should be bestridden by proud arches of stone; these wild prairies should be loaded with wheat; the swamps with rice; the hill-tops should pasture innumerable sheep and cattle.... How much better when the whole land is a garden, and the people have grown up in the bowers of a paradise (Quoted in Merchant 1995: 142).

But even though the word remained the same, the cultural construct that lay behind it changed, especially in the United States, first to mean adventure, then, in the context of the almost entirely mythic American West, to become a defining element of what it meant to be an American (Cronon *et al.* 1992). "Wilderness" became the place where paradise was recovered, "the recovery ... of an original state of innocence and plenty through a return to nature" (Slater 1995: 116).

The evolution of these complex ideas is not unrelated to the evolution of our infrastructures; indeed, it reflects at least in part the reflexivity between foundational infrastructures and culture. To follow up the example in the last paragraph, it is arguable that there is a profound linkage between our dominant technological systems, and the way we think about "nature," perhaps one of the most fundamental cultural constructs, certainly these days (Allenby 2003). Thus, it is fascinating to read about the history of the automobile, and to realize that for many urban populations in the early decades of this century it represented a major means by which they realized "nature." For example, the 1925 Bronx River Parkway from New York City into Westchester County was originally conceived by Frederick Law Olmsted, who designed Central Park, as a "scenic utopia" reflecting an "authentic" American landscape (Gandy 2002). It was planted with over 30,000 trees, each selected to represent those that were, or might have been, indigenous to the area. In this, it was similar to Olmstead's Fens and Riverway project in Boston, which replaced acres of polluted mud flats with carefully engineered and designed "natural" patterns and plantings – or, for that matter, Central Park itself, a supremely engineered statement of what "nature" was for a city determined to ascend into the elite league of world cities. This piece of infrastructure was in fact a highly political statement, inviting one to contemplate "nature" as thus constructed from many angles: conservation of species, reconstruction of (highly idealized) landscapes, statements of power and class in a rapidly developing, immigrant-rich America. Similarly, Robert Moses designed low bridges over his parkway to Jones Beach on Long Island to prevent buses, and thus the lower classes, from accessing it – although, of course, there was plenty of parking for the individual cars of the middle class. One could contemplate how each project explicitly redefined "nature" as an obviously designed and built environment at the time, yet came to be seen by future generations as "natural" in a foundational sense.

On one level, then, the Parkway and similar constructions were familiar civil engineering infrastructure achievements. But on another, more complex, level they were parts of a technological redefinition of basic values to fit changing times. As Gandy (2002: 122) comments, "Nature became simultaneously more distant (framed by the window of a moving car), more accessible (through greater public contact with remote areas), and at the same time more individualized as an aesthetic experience." In other words, automobile technology and its supporting infrastructure created a new, and unique, experience of "nature" – not alone, of course, for the automobile, the Parkway, and other infrastructure were in turn representative of a certain moment in history and culture. And this is not anomalous: "nature" through the window of the automobile becomes "nature" as it appears through the window of the television, still more distant, yet, thanks to the power of "nature" photography, even closer, more emotionally evocative, and more accessible. Nothing is more artificial and contrived than the "nature" program on television – and more culturally potent, for that matter.

And, finally, there is "nature" as it begins to appear across the Internet, in at least three guises. The first, of course, is virtual reality: as bandwidth and processing power grow, "nature" will become entirely a human creation, whether in video games or created experiential landscapes of the future. The second is in complex models interpreting massive data sets: a "nature" that because it is "scientific" will appear more real than the former, but will be just as constructed and ideological. And just as removed, one might add, from whatever really is "out there." And, of course, consumers will increasingly buy their "nature," from geode to cactus garden, not just from the museum or nature store at the mall, but over the Internet: "nature" as e-commerce. Thus, we have come full cycle: infrastructure becomes not just an artifact of a particular time and place, but it becomes the means by which human perceptions of our world are created and focused: the automobile and the Internet are not just increasingly information intensive examples of built infrastructure, but the means by which we perceive the terraformed world around us. The reflexivity with culture, and with "natural" systems, becomes complete.

In summary, IT forms a complicating integrative infrastructure: it is a critical modern infrastructure in itself; it forms an important mechanism by which cultural constructs and critical perceptions evolve; and it is an increasingly complicating and formative component of other infrastructures. The model towards which this understanding leads is one of complex and interconnected networks of networks, some primarily natural, some primarily built, and some primarily cultural and informational. It is not one that we understand as yet.

Technological infrastructure and cultural construct; integrated human and natural systems; regional and global scales; emergent characteristics in natural, economic, technological, and cultural domains – these are the challenges posed by the auto-catalyzing structures of the modern world. Infrastructure at one scale is, as it always has been, an effort in good engineering – good design and operation of artifacts and the constructed systems they are part of. But with the advent of "smart infrastructure" networked across previously isolated systems; the increasing technological, economic, cultural, and social complexity of the world;

and the increase in scale of human activity over the past hundred years (Turner *et al.* 1990; McNeil 2000), we have entered a new era in infrastructure design and operation, the era of earth systems engineering and management. It is far too early to suggest that we know how to do this – but clearly infrastructure systems like modern air transport, ground transport, and the Internet are demonstrating significant emergent characteristics that cannot continue to be ignored. The question, then, is not whether our infrastructure will create an anthropogenic world, for it already has. The question is whether we will react responsibly, rationally, and ethically to that challenge. And that question remains open.

The Anthropocene: Engineering Infrastructure in a Terraformed World

There are a couple of ways to respond to the evolution of the anthropogenic Earth. The first, and by far the most common, is denial. The idea that we are not just moving towards an anthropogenic world, but in fact are already there, is a difficult one for many people, particularly if they view "nature" as sacred and apart from the human (see, for example, *The End of Nature* (McKibben 1989)). But the important point is that we are already in such a world. After all, the entire process of climate change is a belated recognition that, in fact, the accumulated activity of the past millennia or so has resulted in the situation where the human elements dominate the behavior of the carbon cycle, and through that the climate system, in certain important ways. It is premature, but by far less than one might wish, to in fact understand the carbon cycle as a human infrastructure. As genetic engineering builds from the micro scale, and large scale urbanization, land use, and technological change builds from the macro scale, such systems will, in fact, increasingly become human infrastructure. It is important to understand that this statement is intended to be neither grandiose technological optimism nor imply a normative judgment: it is simply an effort to perceive clearly the world as it actually is. Beyond a certain point, self-deception, especially as it leads to increasingly dysfunctional responses, becomes unethical.

What's frightening to many, of course, is the perfectly valid understanding that neither as individuals, nor as institutions, do we know how to manage an Earth where fundamental natural systems become integrated into designed infrastructure – or, in some cases, become human infrastructure. The climate change negotiation process so far is a very primitive first step, and perhaps more illustrative of how far we have to go in our understanding than how far we have come. But in a sense, anybody who designs and operates important infrastructure activities runs up against these same issues. The complexity that infrastructure design and operation increasingly entails arises not just from complicated technology, and the enhanced complexity created as IT and other infrastructures interact and coalesce. Rather, there are difficult cultural dimensions to these technical activities. This requires different ways of engineering. For example, one of the mistakes technologists in general tend to make is to try to solve cultural issues through technological fixes. The reasons are fairly obvious: technologists are trained in, and comfortable with, the technological but not the cultural, so the tendency is to take cultural issues that are extremely difficult and push them into a technological silo. But this really doesn't work in many cases, especially as infrastructure systems at all scales increase in complexity and information content. As a society, we need to develop infrastructure engineers that are as good with culture and politics as they are with materials choices and technical design.

To return to the climate change design problem, consider the failure of the United States and Russia to adopt the Kyoto Treaty reducing carbon dioxide emissions. The language used by most participants in the dialog might lead one to believe that the issue was primarily disagreement over the relevant science. But in fact different parties to the climate change negotiations have very different visions of a desirable future world, and all the scientific data we will ever develop are inadequate to mesh those visions through negotiation. This may be seen, perhaps, by considering the possibility of extracting carbon dioxide at ambient atmospheric concentrations from the atmosphere itself, thereby changing a debate that currently appears to be over carbon dioxide emissions, to a much more profound question: what atmospheric chemistry do you want me to engineer? That question, in turn, directly raises the normative question that in fact lurks, generally unrecognized, under the climate change debate even now: what kind of world do you want? This is a normative, not a scientific question, and scientific data are simply not relevant to the real issue at stake (although they may be very relevant in helping us understand the costs, benefits, and possible paths between here and there).

Addressing the (usually unrecognized) cultural dimensions of these earth systems, engineering and management systems creates a complexity that is compounded by changes in global governance systems. It used to be the case that national governments completely dominated any international issues: they were, indeed, sovereign in both law and reality. But such is no longer the case. It is not so much that national governments have become obsolete, for they surely are not; rather, it is that parallel systems of information transfer, knowledge management, and organizational structure have evolved. Thus, for example, human rights and environmental non-governmental organizations have proliferated in both number and power over the past two decades, disrupting the absolute authority of the nation-state (Sassen 1996; Mathews 1997). The growth of the Internet has created communities of interest that can be rapidly mobilized and dispersed, disrupting decision-making processes that require longer cycles to complete. And increasingly, there is the possibility of single interest organizations dominating a decision process in ways that might not have been possible before. So arises another challenge to infrastructure designers at all scales: IT not only functions as part of infrastructure these days, but also changes the governance context within which infrastructure initiatives are proposed, designed, financed, and executed.

At this level, understanding virtually any infrastructure raises a number of interesting challenges. Among the most apparent is the difficulty of deciding how to bound these complex systems at higher levels to enable analysis and appropriate design. It does not do, after all, to simply claim that everything is connected to everything else; what is necessary is to enable intelligent understanding and management of the system at various levels. Here it is clear that our understanding is still nascent, although a few principles can already be discerned – such as the idea that the question asked defines the relevant system for the answer, as discussed above.

Consider a different case – say, the operation of modern air transport systems. Even starting with the artifact is problematic. Ask the simple question: What is the environmental and social impact of a long distance jet aircraft? If you do the traditional analysis you might say "well the main lifecycle impact of a jet plane is that it uses a certain amount of fuel and produces a certain amount of exhaust." But in fact that technology has enabled the growth of a global tourism industry that has probably had more impact on the biosphere since the European migrations during the Enlightenment. And this is true not just in the obvious sense of burdening desirable locations with increased traffic, but in more subtle senses such as knitting together human population centers, thus enabling more efficient infection by new disease vectors. So should these effects be considered as one contemplates design of a jet aircraft? Most people would probably argue no. Should they be considered as part of the air transport infrastructure? That is a closer question, and much more difficult, for it can be argued that the air transport infrastructure taken as a coherent system is an emergent characteristic of a lot of decisions involving subsystems, and has never been deliberately planned as a whole. In this, the air transport system is similar to many complex infrastructures, such as the Internet, that are clearly human in origin, but just as clearly not planned as integrated systems. And, like many of the infrastructure systems discussed in this volume, it also is enabled, and increasingly defined, by relevant information structures and the IT networks that it, and they, ride on.

At this level, then, we begin to approach the earth systems engineering and management challenge of the modern technological society: we are clearly creating entirely human infrastructures at a level of complexity where their design, function, and longer-term performance and impacts are little researched and poorly understood. But even to recognize this is to understand that both ethics and rationality demand that we embark on the long and complex journey to do so.

Note

1 As discussed in more detail in the rest of this chapter, terms such as "wilderness," "nature," "sustainable development," and the like are cultural constructs: that is, they reflect the particular place and time within which they are used and understood. As such, even the basic language within which we discuss these issues is a function in part of IT infrastructure (Allenby 2002).

References

Abrams, M.H. (1971) *Natural Supernaturalism: Tradition and Revolution in Romantic Literature*, New York: W.W. Norton & Company.

Allenby, B.R. (1992) "Design for Environment: Implementing Industrial Ecology," Ph.D. dissertation, New Brunswick, NJ: Rutgers University.

Allenby, B.R. (1999) *Industrial Ecology: Policy Framework and Implementation*, Upper Saddle River, NJ: Prentice-Hall.

Allenby, B.R. (2002) "Observations on the Philosophic Implications of Earth Systems Engineering and Management," Batten Institute Working Paper, Charlottesville, VA: Batten Institute, Darden Graduate School of Business, University of Virginia.

Allenby, B.R. (2003) "Look! In the window! A cultural construct," *Green Business Letter*, February, p. 8.

Allenby, B.R. and Roitz, J. (2003) "Implementing the Knowledge Economy: The Theory and Practice of Telework," Batten Institute Working Paper, Charlottesville, VA: Batten Institute, Darden Graduate School of Business, University of Virginia.

American Electronics Association (AEA) (1992) *The Hows and Whys of Design for the Environment: A Primer for Members of the American Electronics Association*, Washington, D.C.: AEA.

Barabasi, A. (2002) *Linked: The New Science of Networks*, Cambridge, MA: Perseus Publishing.

Cronon, W. (ed.) (1996) *Uncommon Ground: Rethinking the Human Place in Nature*, New York: W.W. Norton & Company.

Cronon, W., Miles, G. and Gitlin, J. (eds) (1992) *Under an Open Sky: Rethinking America's Western Past*, New York: W.W. Norton & Company.

Gandy, M. (2002) *Concrete and Clay: Reworking Nature in New York City*, Cambridge: The MIT Press.

Heidegger, M. (1977) Original essays 1952–1962, collected 1977. *The Question Concerning Technology and Other Essays*, W. Lovitt, translation, New York: Harper Torchbooks.

Mathews, J.T. (1997) "Power shift," *Foreign Affairs* 76(1): 50–66.

McKibben, B. (1989) *The End of Nature*, New York: Random House.

McNeill, J.R. (2000) *Something New Under the Sun*, New York: W.W. Norton & Company.

Merchant, C. (1995) "Reinventing Eden: Western culture as a recovery narrative," in Cronon, W. (ed.) *Uncommon Ground: Rethinking the Human Place in Nature*, New York: W.W. Norton & Company, pp. 132–59.

Nature (2003) "Welcome to the Anthropocene," *Nature*, August 424: 709.

Sagoff, M. (1996 first published, 1988) *The Economy of the Earth*, Cambridge: Cambridge University Press.

Sassen, S. (1996) *Losing control? Sovereignty in an Age of Globalization*, New York: Columbia University Press.

Slater, C. (1995) "Amazonia as Edenic narrative," in Cronon, W. (ed.) *Uncommon Ground: Rethinking the Human Place in Nature*, New York: W.W. Norton & Company, pp. 114–31.

Turner, B.L., Clark, W.C., Kates, R.W., Richards, J.F., Mathews, J.T. and Meyer, W.B. (eds) (1990) *The Earth as Transformed by Human Action*, Cambridge: Cambridge University Press.

Vitousek, P.M., Mooney, H.A. Lubchenco, J. and Melillo, J.M. (1997) "Human domination of Earth's ecosystems," *Science* 277: 494–9.

WECD (1987) The World Commission on Environment and Development, (a.k.a. the Brundtland Commission) *Our Common Future*, Oxford: Oxford University Press.

Part IV
Conclusion

CHAPTER THIRTEEN

Themes and New Directions

Thomas Horan and Rae Zimmerman

Introduction: Information and Detection

The world was a simpler place when Dr John Snow discovered how a single water well accounted for London's cholera outbreak in 1854. While almost 500 had died in early September of that year, he was able to operate in a detective-like fashion to investigate the environmental conditions, social context and spatial distributions associated with this outbreak and hone in on the Broad Street well as the infecting infrastructure (Snow 1855).

The manner in which Dr Snow went about his work during that fall in Soho remains valuable today. Strong analytic skills are needed to detect, monitor, analyze, and correct the functioning of infrastructure systems. But the world has become amazingly more complex; for example, as pointed out in Chapter 5 an estimated 1.3 million waterborne disease cases occur annually in the US alone (WaterWorld 2002; Payment *et al.* 1997). Moreover, due to environmental policy advancements and regulations (to avoid outbreaks such as cholera) in the US and worldwide, a variety of delivery and monitoring systems are being deployed to help safeguard the quality of this vital natural resource. Similar technologies surround other civil and environmental infrastructure systems.

Clearly, a major infrastructure challenge of this digital age is to harness the information produced by digital infrastructures to enhance the social, economic, and environmental condition of our "global village." This final chapter draws upon the insights from the previous sections as well as related literature to consider themes and directions for digital infrastructures.

Recognizing Nodes of Interdependence

As introduced in Chapter 1, a fundamental tenet of this volume is the converging nature of civil and environmental infrastructures and the digital infrastructures that support them. The book chapters probe the interconnections that occur between and across systems. For example, Amin summarizes the reliance of electricity on the telecommunications system for monitoring and control of vast networks; Zimmerman demonstrates how water systems are reliant on IT to sustain ever-increasing service areas, to coordinate electricity and transportation for the pumping, treatment and transport of water and wastewater; and to meet greater stringency of standards. Cluett presents an extensive array of IT-dependent systems in the transportation sector aimed at improving safety and mobility.

The technology and economic chapters provide further documentation on the extent to which IT has become enmeshed into infrastructure systems. In terms of technology, O'Brien and Soibelman describe the important role that IT now plays in sensing, monitoring and controlling infrastructure – a role that will become only more complex and critical with the onset of nanotechnology capabilities. Infrastructures of tomorrow could very likely be covered with "smart dust" that will provide an unprecedented amount of data on the infrastructure and how it is performing (Doherty *et al.* 2001).

The economic analysis by Henry and Dumagan suggests, however, that one need not wait for smart dust to establish strong linkages between IT and infrastructures. Their analysis found fully one-third of all IT investment was made by infrastructure industries. Specifically, between 1996 and 2001, infrastructure industries, which together accounted for eight to nine percent of the nation's Gross Domestic Product (GDP), owned about one-third of all capital stock in IT equipment and software. However, similar to the productivity paradox of other sectoral uses of IT, the authors found similar circumstances surrounding digital infrastructure investment. They write, "despite this heavy investment in digital products, in general, the productivity growth in infrastructure industries was not dramatic and did not contribute significantly to the overall labor productivity acceleration that happened in the late 1990s." Whether the advent of new systems based on nanotechnology will improve this productivity remains to be seen. But one fact is certain, civil and environmental infrastructures will continue to be major targets for IT investment and development.

These sectoral, technological, and economic analyses all lead to the same question: How can promising IT systems be devised to reliably deliver the information needed to maintain our increasingly complex civil and environmental infrastructures? To address this question, the infrastructure industry (including private, governmental and nonprofits) has begun to embrace a strong technology research and development orientation to civil and environmental infrastructure. The digital infrastructure has become critical to the functioning of cities and towns throughout the developed world, and the developing world is not far behind as Restrepo points out. The technology research activities (in electricity) presented by Amin are but an illustration of the dedicated effort that is required to make continued gains in using IT to support civil and environmental infrastructures.

While digital infrastructures, by definition, have a strong technical element, this is not to understate the management and policy challenges that surround these systems. The nature of interactions across infrastructure systems and the environment they operate in remains an "emergent" phenomenon. Horan and Sparrow suggest that complexity theory provides a useful framework to understand how digital infrastructure systems operate and respond to adaptive management techniques. Allenby builds on and extends this complexity concept by observing that digital infrastructures can fundamentally change our perceptions of both the physical, human and natural systems. The observations by volume authors echo the general observations of scores of prominent scholars, who have observed similar dynamics across a range of technologically embedded circumstances (e.g. Anderson 1999, Arthur 1994, Perrow 1984, Simon 1996, Winner 1977). A common theme throughout this work is the need for and value of redundant and resilient systems and the important role that IT plays in achieving these system characteristics.

In the following sections, we outline several areas where attention is warranted to further the effective integration of digital infrastructures into civil and environmental systems. The first is extreme events, which provide important conditions under which complex systems can be examined.

Understanding Extreme Events

Two decades ago Charles Perrow (1984) provided a lucent theoretical basis for understanding extreme events and dealing with their inevitability. Events since then have only strengthened his belief in the importance of the concept (Perrow 1999). In a paper prepared for a conference on Extreme Events, he proclaimed:

> Extreme events are critical determinants of the evolution and character of many – perhaps most – natural and human-influenced systems. Conversely, it is no overstatement to suggest that humanity's future will be shaped by its capacity to anticipate, prepare for, respond to, and, when possible, even prevent extreme events. From such a perspective, extreme events emerge as a powerful focus for organizing research activities that can advance scientific knowledge and directly benefit society (Perrow 2000: 2).

Unfortunately and even tragically, the US has recently begun to experience terrorist activities that have plagued other parts of the world for decades if not for centuries. Many chapters acknowledge that September 11, 2001 remains a pivotal day for the US, as viewed from the infrastructure perspective. The Moss and Townsend chapter directly addresses the IT dimension to this day and how the effects of the disruptions of IT were felt throughout many different infrastructure services. Their chapter notes how the collapse of the World Trade Center stressed the telecommunications infrastructure to an extent not heretofore experienced in New York City, or indeed in the country. They write, for example, that: "during the morning of September 11, fewer than 1 in 20 mobile calls were connected in New York City. Since no system was in place to prioritize calls for emergency workers, response was confused and uncoordinated throughout much of the day." Other chapters presented examples from transportation electricity, water, and (emergency) telecommunications that underscored how accidents can adversely affect infrastructures and the role that they can play in both propagating and resolving system failures.

Many of these examples either explicitly or implicitly note the importance of redundancy in both the physical and IT network. The importance of network redundancy was observed at a conference convened to study the impacts of September 11 (MCEER 2002). A principal finding of the conference was that New York City was able to restore critical services and initiate early recovery activities relatively quickly due to the inherent redundancy of many of its physical and institutional infrastructures. Other cities may not be so fortunate,

> Less robust systems, or infrastructure systems in less highly resilient cities, would likely not have fared as well. Infrastructure systems in major urban areas are inherently interconnected and vulnerable to complex system failures. We do not fully understand all of the vulnerabilities that can cause losses to proliferate and hinder recovery efforts (MCEER 2002).

This lack of understanding of how extreme events have affected critical infrastructure has become a concern in the US and internationally. In the US, the National Strategy for the Physical Protection of Critical Infrastructure and Key Assets (White House 2003) recognizes both the importance and interdependence of physical infrastructures systems (including their digital subsystems): "The facilities, systems, and functions that comprise our critical infrastructure are highly sophisticated and complex. *They include human assets and physical and cyber systems that work together in processes that are highly interdependent*" (White House 2003: viii, emphasis added). The plan goes on to highlight the need for a robust research program involving standards development, modeling and simulation and coordinated deployment across agencies. As highlighted in the quotation above, the plan recognizes the key role IT systems play in sensing, monitoring, and controlling critical infrastructures. This volume has served to underscore the extent to which IT is already impacting civil and environmental infrastructures and highlights the need to focus on cross-infrastructure dependencies from both technology and social science perspectives.

At the international level, concern has grown about the ability to protect critical infrastructures. In a paper prepared for an International Telecommunications Union (ITU) conference on the subject, Goodman *et al.* (2002) writes:

> Communication has become the lifeblood of modern societies. The rise of e-commerce has made these networks responsible for a growing share of national wealth and hopes for greater prosperity. Furthermore, in some nations, pre-existing critical services have come to rely on electronic networks – emergency services, navigation systems for shipping and air traffic, electric power grids, and water control systems. While these

> dependencies vary from nation to nation, nearly all nations already or will in the future depend on these critical network infrastructures (Goodman *et al.* 2002: 2).

Goodman, *et al.* outline several areas for international cooperation in (cyber) security; and these have implications for future directions for digital infrastructure protection. These dimensions include: international standards (e.g. wireless Wired Equivalent Privacy (WEP) encryption standard), information sharing (e.g. about critical infrastructure assaults), halting attacks in progress (e.g. real-time coordination to protect critical infrastructures), coordinating legal systems (e.g. a set of international rules regarding prosecution for critical infrastructure attacks), and providing assistance to developing nations (e.g. technical assistance is setting up security systems). These dimensions provide a useful starting point for considering future international linkages and cooperation in the digital infrastructure arena.

Designing Systems for Users

While nations and their agencies can provide policies and regulations to help advance and protect digital infrastructures, in the end, digital infrastructures are operated and used by people. The human factor remains a critical link and, therefore, attention is needed to design systems that can be operated and used during normal and extreme events. In their chapter on technology, O'Brien and Soibelman note that the range of actors involved in the information and physical infrastructures pose a considerable challenge to the successful implementation of such systems. They point out the particular challenge of developing and conveying information that can be comprehended and acted upon by individual users.

Perrow (2000) confirms the importance of usable information, particularly in times of crises.

> The key issue is how to integrate and present information in ways that are useful to decision makers. Different decision makers operate under different sets of rules and capabilities. Following recent catastrophic flooding in Mozambique, efforts to transmit digital satellite data to emergency planners paralyzed the nation's computer networks, creating a second (albeit minor) extreme event. During the 1997 Red River Flood in North Dakota, copious observational data about river conditions were available, but weather forecasters and local decision makers communicated poorly, magnifying the scale of the disaster (Perrow 2000: 8–9).

To Horan and Sparrow, the issue is one of design in the broadest sense of the term: in their view the design of digital infrastructures needs to include elements relating to users, systems, organizations, as well as the social-political context. Managers of these systems need to develop a holistic view of an entire systems performance, have skills in inter-organizational and dynamic management, and be trained in a setting that provides contextualized problem-solving.

The organizational change challenge is echoed by Cluett's conclusion for transportation systems. He notes that a key to such change is creating an organizational culture that is comfortable with the infusion of high technology, a more rapid pace of decision-making, and a greater degree of customer orientation and responsiveness. He notes that a related cultural change that is needed is increased openness and willingness to share information within and among agencies and across jurisdictions, with more attention to joint problem-solving.

Surrounding users and institutions is the policy architecture and its design. The chapter by Gifford highlights the important role of public policy in affecting the nature and range of IT influence on infrastructures. He cites a number of concerns that policy must deal with, ranging from economic efficiency to privacy. Moreover, he provides an analysis of the policy levers at the disposal of policy makers in the form of support for research and development, deployment and field-testing, system architecture and standards, financial incentives, and planning and procurement requirements. Beatley takes the issue of design to the physical level in terms of how IT can enhance green design of buildings and neighborhoods. Drawing on a wide range of implementations throughout Europe, Asia, and the US, he demonstrates that "smart and green" design is achievable today but depends on integrated and thoughtful decisions by myriad stakeholders.

Taken together, these are all examples of policy approaches that can be taken at the federal, state, and local level to support the integration of IT into civil and environmental infrastructures. The actual choice of such levers represents the policy design dimension that would accompany appropriate managerial, information system, and physical infrastructure design elements. It is important to emphasize that such policy decisions need to be made with an eye to the equity implication of digital infrastructures. In the US, issues of social equity have been pervasive in public policy at least since the civil rights movement of the early 1960s. Since that time, equity issues have diffused into many specific policy arenas, and in particular, in the provision of public infrastructure services. Equity concerns of minorities and the economically disadvantaged have surrounded infrastructure decisions, such as how the location-decision for hazardous waste facilities could impact the environmental, social, and economic viability of low income neighborhoods (United Church of Christ 1987; US General Accounting Office 1995). In transportation, equity concerns have surrounded the location of transport facilities (Bullard and Johnson 1997), and, more recently, the pricing of those facilities and services (Taylor, Wachs, *et al.* 1995). As the use of IT increased dramatically, equity concerns soon spread to this area as well, popularly known as the "digital divide" (US Department of Commerce 2000). Extreme events exacerbate equity problems further, and when natural disasters such as floods and earthquakes occur, infrastructure is a common casualty with poorer people and minorities having less of an ability to cope than those with greater access to resources. Internationally, equity issues are somewhat similar to those in the US, however, often at issue are the timing and stringency of constraints posed on developing countries that had not been imposed on developed countries during their era of industrial growth. This particularly arises in the areas of transportation and energy.

Thus, though issues of equity arise similarly but separately for IT and other infrastructure, one can surmise that the combination of the two – where IT is used to enable infrastructure – could compound equity problems unless deliberate and effective policy action is taken. Little is known as to whether this is or will be the case for digital infrastructure, but care must be taken to ensure that equity considerations are thoughtfully integrated into IT implementation decisions.

Addressing the Global Dimension

Marshall McLuhan envisioned the emergence of the "global village" in terms of ubiquitous sensing of events: "Today, after more than a century of electric technology, we have extended our central nervous system in a global embrace, abolishing both space and time as far as our planet is concerned" (McLuhan 1964: 3).

In this volume, O'Brien and Soibelman confront the practical challenge of implementing this global nervous system. Drawing on the work on Aktan, they note the sensing infrastructure associated with one bridge – 485 sensors with capabilities to monitor traffic, weather, and structural response. They observe that "as each channel provides real-time sensor data, it is clear that no human could effectively monitor the data to make intelligent decisions about structural performance and associated activities such as identifying maintenance needs."

Despite these challenges, the role of IT in guiding global development is significant. In his article, "The New Urban Planning," Jeffrey Sachs (2003) outlines a five step process for bringing developing cities into the twenty-first century. Technology of course, is one such dimension:

> Finally, a global strategy should include a globally-networked research strategy as well. Satellite mapping and imagery, remote sensing and geographic information system (GIS) data can provide great insights into the dynamics of urban areas over time. The research agenda is fascinating and crucial for global development, and it will attract leading minds worldwide. It's time to give a major push forward to research on urban sustainable development in the poor countries of the world.

Restrepo provides a sobering profile of the needs that confront such developing countries. While developed regions of the world must integrate digital technologies into relatively mature infrastructure systems, the situation for developing regions is quite different. In these (developing) regions, traditional and digital infrastructures have to be advanced concurrently. Moreover, consideration needs to be given to the cost-saving opportunity to leap-frog traditional paths of infrastructure development, so as to take advantage of downward cost trends and/or new technological advances. For example, the expected growth in automobile emissions in developing countries represents a serious global environment challenge, yet integrated approaches involving technology are just being formulated. Both China and Mexico have undertaken integrated reviews of their transport sector, and findings from these reviews highlight the role of technology (including but certainly not limited to IT related efforts) in assisting in the monitoring and reduction of automobile-based emissions (Molina and Molina 2002; National Academy of Sciences 2003).

For many regions, it is the electricity infrastructure that presents the most pressing challenge for development. As Restrepo notes,

> in terms of efficiency in electricity systems around the world, developing countries seem to be at a loss; their grids have much higher transmission and distribution losses as a percentage of total output than those in developed countries. In Nigeria these losses amounted to over 30 percent of the total in 1996, and Kenya, Pakistan and India had losses greater than 15 percent of total output. At the other end of the scale, Japan and Singapore had losses as small as four percent. Although in most of the countries shown these losses were reduced between 1980 and 1996, in Mexico and Brazil they increased.

The same problem exists in water Zimmerman cites sources that indicate much larger water loss rates in developing countries than in developed countries.

In his analysis of the electricity infrastructure, Amin reminds us of the problems that have been caused by its shortage in developed countries. While the US media gives extensive coverage to North America, there is no doubt that blackouts are a global phenomenon – in 2003 alone there were blackouts in the UK south London on August 28, in southern Sweden and parts of Denmark on September 23, and Italy on September 28, affecting a total of approximately 112 million persons. He also cites earlier ones in France (December 1978) from a voltage collapse, and destruction of high-voltage transmission towers in France (December 1999).

In short, civil and environmental systems have global dimensions that need to be addressed and IT will increasingly play a key role in ensuring the reliability of these systems. Global attention and leadership is needed, however, to ensure that developing as well as developed countries have sufficient technical expertise and resources to adequately deploy and maintain these systems.

Providing Leadership for a Sustainable Future

In his collection of essays on *Conservation in the Internet Age*, James Levitt (2002) notes the historical symmetry between industry titans of an age and the environmental conditions of that age. The Rockefeller family benefited immensely from the oil surge from automobile use, yet went on to establish a major conservation foundation in the United States, including playing a key role in the establishment of the national parks. In this sense, it is perhaps fitting that the titans of the information and communications industry – Gordon Moore, Bill Gates, Ted Turner – are now filling a philanthropic vacuum with regard to the global challenges of this age.

This IT leadership has focused on the global environmental and public health arenas. As an example of the role of digital technology in the international environmental arena, the Gordon and Betty Moore foundation is supporting an unprecedented initiative to monitor and protect the remaining rain forests in the world. A key to this initiative is establishing monitoring stations in 25 biodiversity hotspots. These hotspots amount to just 1.4 percent of the land surface of the planet, yet support nearly 60 percent of the world's plant, bird, mammal, reptile, and amphibian species (Conservation International 2003). This monitoring effort will include use of satellite, aerial, and field observations to characterize and monitor the impacts of human activities on biodiversity in the hotspots. According to the Center for Applied Biodiversity Science (2003) "integrating this new generation of space and airborne remote sensing instruments with comprehensive databases on social, economic, political, and legal factors enables us to better understand the relationships between the biophysical environment and patterns of human use."

While major philanthropic (not to mention governmental) initiatives are sorely needed to address global conditions through innovative use of technology, by far the majority of work will be carried on by professionals and citizens doing their part to improve their local communities. The final but perhaps most important theme running through the volume is the need to educate and train professionals and citizens alike in thoughtful use of technology to improve local infrastructure services and conditions. Each one of the chapters pointed to the need for building such capabilities if IT is indeed to enable infrastructure at a very practical level. At the present time, knowledge and communication barriers exist between the two areas often to the point where infrastructure professionals are not able to hire and manage experts to integrate IT into their systems. As noted at the outset of this volume, IT represents a two-edged sword – one edge providing solutions, the other creating problems. The many cases summarized in the volume's chapters reveal that the wielder of this sword is most often professional and citizen decision-makers and users. This is where the real leadership will reside and, consequently, where digital infrastructure education and training efforts need to focus.

Educating a New Generation

Indeed, one of the strongest messages conveyed by each of the chapters is the need for education programs to enhance the quality of infrastructure systems. First, as pointed out by Henry and Dumagan, changes in the workforce in IT industries (such as retirements) and the extensive demands of the infrastructure sectors for those services creates a need for educational programs for new workers entering this area. Second, all the infrastructure sector chapters identify a need for educating infrastructure professionals in the IT area so that these professionals can effectively manage technical systems, personnel and contractors, and to adapt to the rapidly changing conditions characteristic of IT infrastructure. Green initiatives as described by both Allenby and Beatley emphasize that education of the public and building professionals is needed to over-come many of the obstacles to enhancing such initiatives through the use of IT. Finally, the management and policy chapters emphasize the need for interdisciplinary education to train the more technologically facile "new infrastructure professional." In short, a smart infrastructure requires a smart workforce.

Conclusion

This volume examines an emerging set of systems and approaches for creating an IT-enhanced civil and environmental infrastructure that can improve local as well as global conditions. But as Allenby notes, much is still unknown:

> IT forms a complicating integrative infrastructure: it is a critical modern infrastructure in itself; it forms an important mechanism by which cultural constructs and critical perceptions evolve; and it is an increasingly complicating and formative component of other infrastructures. The model towards which this understanding leads is one of complex and interconnected networks of networks, some primarily natural, some primarily built, and some primarily cultural and informational. It is not one that we understand as yet.

We concur. However, it is our belief that the contributors of this volume chart a promising direction for improving our understanding of such complex digital infrastructures and, importantly, provide valuable insight into how these systems can be deployed to improve social, environmental, economic, and security conditions. If there is a bottom line to the volume, it is this: the next generation of infrastructure systems needs to be fully converged with information technology, needs to integrate information technology into its surrounding social, organizational, and policy systems, needs to use information technology to aid in protecting critical resources under adverse conditions, and needs to consider how information technology can assist in addressing local, regional, national and even global concerns. It may be an enormous task but with proper research, testing, training and education, solutions are quite possible.

References

Anderson, P. (1999) "Complexity Theory and Organization Science," *Organization Science*, 10(3): 216–32.

Arthur, B. (1994) *Increasing Returns and Path Dependency in the Economy*, Ann Arbor: University of Michigan Press.

Bullard, R.D. and Johnson, G.S. (1997) *Just Transportation. Dismantling Race and Class Barriers to Mobility*, Gabriola Island, BC: New Society Publishers.

Center for Applied Biodiversity Science (CABS) (2003) Hotspots Summary, Online. Available at: www.biodiversityscience.org/xp/CABS/research/hotspots/hsconcept.xml (accessed January 15, 2004).

Conservation International (2003) Biodiversity Hotspots, Online. Available at: www.biodiversityhotspots.org/xp/Hotspots (accessed January 15, 2004).

Doherty, L., Warneke, B., Boser, B. and Pister, K. (2001) "Energy and Performance Considerations for Smart Dust," *International Journal of Parallel and Distributed Sensor Networks*, 4(3): 121–33.

Goodman, S., Hassebroek, P., King, D. and Ozment. A. (2002) "International Coordination to Increase the Security of Critical Network Infrastructures," prepared for *Workshop on Creating Trust in Critical Network Infrastructures*, held in Seoul, Republic of Korea, from May 20 to 22, 2002.

Levitt, J. (2002) *Conservation in the Internet Age*, Washington, D.C.: Island Press.

McLuhan, M. (1964) *Understanding Media*, Cambridge, MA: MIT Press.

Molina, L. and Molina, R. (2002) *Air Quality in the Mexico Megacity: An Integrated Assessment*, New York: Kluwer Academic Press.

Multidisciplinary Center for Earthquake Engineering Research (MCEER) (2002) "Lessons from the World Trade Center Terrorist Attack Management of Complex Civil Emergencies and Terrorism-Resistant Civil Engineering Design," New York City, June 24–25.

National Academy of Sciences (2003) *Personal Cars and China*, Washington, D.C.: National Academy Press.

Payment, P., Siemiatycke, J., Richardson, L., Renaud, G., Franco, E. and Prevost, M. (1997) "A Prospective Epidemiologic Study of the Gastrointestinal Health effects Due to the Consumption of Water," *International Journal of Environmental Health Research*, March, pp. 5–32.

Perrow, C. (1984) *Normal Accidents: Living with High-Risk Technologies*, New York: Basic Books.

Perrow, C. (1999) *Normal Accidents: Living with High-Risk Technologies, Second Edition*, Princeton, NJ: Princeton University Press.

Perrow, C. (2000) "Extreme Events: A Framework for Organizing, Integrating and Ensuring the Public Value of Research," Paper prepared for *Extreme Events: Developing a Research Agenda for the 21st Century*, Boulder, CO, July 7–9.

Sachs, J. (2003) "The New Urban Planning," *Development Outreach*, Washington, D.C.: World Bank Institute, November.

Simon, H.A. (1996) *The Sciences of the Artificial*, 3rd edn, Cambridge, MA: MIT Press.

Snow, J. (1855) *On the Mode of Communication of Cholera*, London: John Churchill, New Burlington Street, England.

Taylor, B.D., Wachs, M. *et al.* (1995) "Variations in Fare Payment and Public Subsidy by Race and Ethnicity: An Examination of the Los Angeles Metropolitan Transportation Authority," Los Angeles, CA: School of Public Policy and Social Research, January 5.

United Church of Christ, Commission on Racial Justice (1987) *Toxic Wastes and Race in the United States*, New York: United Church of Christ.

US Department of Commerce (2000) *Falling Through the Net: Toward Digital Inclusion*. A Report on Americans' Access to Technology Tools, Washington, D.C.

US General Accounting Office (1995) *Hazardous and Non-hazardous Waste: Demographics of People Living Near Waste Facilities*, Washington, D.C.: US GAO, June 13. GAO/RCED-

95-84; *Ten Studies on Demographics Near Waste Facilities*, Washington, D.C.: US GAO, June 13. GAO/RCED-95-158R.

WaterWorld (2002) "Survey Results Highlight Infrastructure Need," *WaterWorld*, December, Online. Available at: http://ww.pennnet.com/articles/article_display.cfm?Section=ARCHI&Subsection=Display&ARTICLE_ID=164883&KEYWORD=SCADA.

White House (2003) *The National Strategy for Physical Protection of Critical Infrastructures and Key Assets*, Washington, D.C., February.

Winner, L. (1977) *Autonomous Technology: Technics Out of Control as A Theme in Political Thought*, Cambridge, MA: Riverside Press.

INDEX

Page numbers in *italics* represent Tables. Page numbers in **bold** represent Figures. Page numbers followed by 'n' represent endnotes.